Geometry and Topology for Mesh Generation

The book combines topics in mathematics (geometry and topology), computer science (algorithms), and engineering (mesh generation). The original motivation for these topics was the difficulty faced (both conceptually and in technical execution) in any attempt to combine elements of combinatorial and numerical algorithms. Mesh generation is a topic in which a meaningful combination of these different approaches to problem solving is inevitable. The book develops methods from both areas that are amenable to combination and explains recent breakthrough solutions to meshing that fit into this category.

The book is an ideal graduate text for courses on mesh generation. The specific material is selected by giving preference to topics that are elementary, attractive, useful, and interesting and that lend themselves to teaching.

Herbert Edelsbrunner is Arts and Sciences Professor of Computer Science at Duke University. He was the winner of the 1991 Waterman award from the National Science Foundation and is the founder and director of Raindrop Geomagic, a 3-D modeling company.

T0296487

**CAMBRIDGE MONOGRAPHS ON
APPLIED AND COMPUTATIONAL
MATHEMATICS**

Series Editors
P. G. CIARLET, A. ISERLES, R. V. KOHN, M. H. WRIGHT

6   Geometry and Topology for Mesh Generation

The *Cambridge Monographs on Applied and Computational Mathematics* reflects the crucial role of mathematical and computational techniques in contemporary science. The series publishes expositions on all aspects of applicable and numerical mathematics, with an emphasis on new developments in this fast-moving area of research.

State-of-the-art methods and algorithms as well as modern mathematical descriptions of physical and mechanical ideas are presented in a manner suited to graduate research students and professionals alike. Sound pedagogical presentation is a prerequisite. It is intended that books in the series will serve to inform a new generation of researchers.

Geometry and Topology for Mesh Generation

HERBERT EDELSBRUNNER
Duke University

CAMBRIDGE
UNIVERSITY PRESS

CAMBRIDGE UNIVERSITY PRESS
Cambridge, New York, Melbourne, Madrid, Cape Town, Singapore,
São Paulo, Delhi, Dubai, Tokyo, Mexico City

Cambridge University Press
The Edinburgh Building, Cambridge CB2 8RU, UK

Published in the United States of America by Cambridge University Press, New York

www.cambridge.org
Information on this title: www.cambridge.org/9780521793094

© Cambridge University Press 2001

First published 2001
First paperback edition 2006

A catalogue record for this publication is available from the British Library

Library of Congress Cataloguing in Publication Data
Edelsbrunner, Herbert.
Geometry and topology for mesh generation / Herbert Edelsbrunner.
p. cm. – (Cambridge monographs on applied and computational mathematics ; 7)
Includes bibliographical references and indexes.
ISBN 0-521-79309-2
1. Numerical grid generation (Numerical analysis) 2. Geometry. 3. Topology.
I. Title. II. Series.
QA377 .E36 2001
519.4 – dc21 00-046777

ISBN 978-0-521-79309-4 Hardback

To Ping, Xixi, and Daniel

Contents

Preface

The title of this book promises a discussion of topics in geometry and topology applied to grid or mesh generation. To generate meshes we need algorithms, the subject that provides the glue for our various investigations. However, I make no attempt to cover the breadth of computational geometry. Quite to the contrary, I seek out the subarea relevant to mesh generation, and I enrich that material with concepts from combinatorial topology and a modest amount of numerical analysis. To preserve the focus, I limit attention to meshes composed of triangles and tetrahedra. The economy in breadth permits a coherent and locally self-contained treatment of all topics. My choices are guided by stylistic concerns aimed at exposing ideas and limiting the amount of technical detail.

This book is based on notes I developed while teaching graduate courses at the University of Illinois at Urbana-Champaign and Duke University. The organization into chapters, sections, exercises, and open problems reflects the teaching style I practiced in these courses. Each chapter but the last develops a major topic and is worth about 2 weeks of teaching. Some of the topics are closely related and others are independent. The chapters are divided into sections; each section corresponds to a lecture of about 75 minutes. I believe in an approach to research that complements knowing what is known with knowing what is not known. I therefore recommend spending time in each lecture to discuss one of the open problems collected in the last chapter.

Chapter 1 is devoted to *Delaunay triangulations* in the plane. We learn what they are and how we can write algorithms to construct them. Although triangulations are inherently combinatorial concepts, we need to answer numerical questions about the relative position of data points. The apparent conflict between logical consistency and numerical approximation is resolved with the help of exact arithmetic and symbolic perturbation. Chapter 2 studies *triangle meshes*, and in particular, the most popular type, which are Delaunay triangulations. The reasons for this popularity are fast algorithms and nice structural properties. In the mesh generation context, Delaunay triangulations are used to represent pieces of a continuous space in a way that supports numerical algorithms computing properties of that space. Such representations are obtained by complementing combinatorial algorithms with numerical point placement mechanisms.

The move from two to three and possibly higher dimensions greatly benefits from precise and concise language. Chapter 3 introduces such language developed within the area of *combinatorial topology*. This relatively old field of mathematics studies the topology of spaces constructed of linear pieces. Chapter 4 puts the language of combinatorial topology to use in our study of *surface simplification*. Given a finely triangulated surface in space, we ask for a coarser triangulation that represents, more or less, the same surface. The need to suppress and compress information through simplification is universal and every bit as strong in the visual arts as in our general quest for understanding.

Chapter 5 generalizes two-dimensional Delaunay triangulations to three-dimensional *Delaunay tetrahedrizations*. Many of the nice properties that hold in two dimensions extend to three dimensions, but some do not. In general, things are more complicated, and a disciplined and formal way of thinking is more important than it is in the plane, where our intuition is often correct. Chapter 6 studies *tetrahedron meshes*, and in particular, the most popular type, which are Delaunay tetrahedrizations. As in two dimensions, the popularity is based on fast algorithms and nice structural properties. While in a sense the shape of Delaunay triangles is as good as it can be for given data points, we need additional methods to eliminate flat tetrahedra and thus improve the quality of the Delaunay tetrahedrization.

Chapter 7 collects 23 problems or questions that, to the best of my knowledge, are open at this time. There is one open problem per section. I make an effort to state each problem in a concise and unambiguous manner and to mention interesting partial results along with general background and motivation.

Herbert Edelsbrunner
Durham, North Carolina, May 2000

1

Delaunay triangulations

The four sections in this chapter focus on Delaunay triangulations for finite point sets in the plane. Section 1.1 introduces the Delaunay triangulation as the dual of the Voronoi diagram. Section 1.2 describes an algorithm that constructs the Delaunay triangulation as a sequence of edge flips. Although the running time of the algorithm is not the best possible, the fact that it halts and is correct allows us to deduce nontrivial structural properties about Delaunay triangulations in the plane. Section 1.3 gives an incremental algorithm whose randomized running time is the best possible. The implementation of a geometric algorithm is generally a challenging task, and the algorithms in Sections 1.2 and 1.3 are no exceptions. Section 1.4 discusses the use of exact arithmetic and symbolic perturbation to implement the numerical aspects with algebraic tools.

1.1 Voronoi and Delaunay

This section introduces Delaunay triangulations as duals of Voronoi diagrams. It discusses the role of general position in the definition and explains some of the basic properties of Delaunay triangulations.

Voronoi diagrams

Given a finite set of points in the plane, the idea is to assign to each point a region of influence in such a way that the regions decompose the plane. To describe a specific way to do that, let $S \subseteq \mathbb{R}^2$ be a set of n points and define the *Voronoi region* of $p \in S$ as the set of points $x \in \mathbb{R}^2$ that are at least as close to p as to any other point in S; that is,

$$V_p = \{x \in \mathbb{R}^2 \mid \|x - p\| \leq \|x - q\|, \; \forall q \in S\}.$$

1

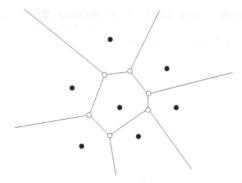

Figure 1.1. Seven points define the same number of Voronoi regions. One of the regions is bounded because the defining point is completely surrounded by the others.

This definition is illustrated in Figure 1.1. Consider the half-plane of points at least as close to p as to q: $H_{pq} = \{x \in \mathbb{R}^2 \mid \|x - p\| \leq \|x - q\|\}$. The Voronoi region of p is the intersection of half-planes H_{pq}, for all $q \in S - \{p\}$. It follows that V_p is a convex polygonal region, possibly unbounded, with at most $n - 1$ edges.

Each point $x \in \mathbb{R}^2$ has at least one nearest point in S, so it lies in at least one Voronoi region. It follows that the Voronoi regions cover the entire plane. Two Voronoi regions lie on opposite sides of the perpendicular bisector separating the two generating points. It follows that Voronoi regions do not share interior points, and if a point x belongs to two Voronoi regions, then it lies on the bisector of the two generators. The Voronoi regions together with their shared edges and vertices form the *Voronoi diagram* of S.

Delaunay triangulation

We get a dual diagram if we draw a straight *Delaunay edge* connecting points $p, q \in S$ if and only if their Voronoi regions intersect along a common line segment; see Figure 1.2. In general, the Delaunay edges decompose the convex hull of S into triangular regions, which are referred to as *Delaunay triangles*.

To count the Delaunay edges we use some results on *planar graphs*, defined by the property that their edges can be drawn in the plane without crossing. It is true that no two Delaunay edges cross each other, but to avoid an argument, we draw each Delaunay edge from one endpoint straight to the midpoint of the shared Voronoi edge and then straight to the other endpoint. Now it is obvious that no two of these edges cross. With the use of Euler's relation, it can be shown that a planar graph with $n \geq 3$ vertices has at most $3n - 6$ edges and at most $2n - 4$ faces. The same bounds hold for the number of Delaunay edges

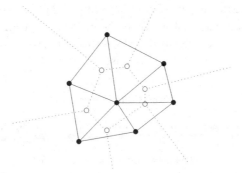

Figure 1.2. The Voronoi edges are dotted and the dual Delaunay edges are solid.

and triangles. There is a bijection between the Voronoi edges and the Delaunay edges, so $3n - 6$ is also an upper bound on the number of Voronoi edges. Similarly, $2n - 4$ is an upper bound on the number of Voronoi vertices.

Degeneracy

There is an ambiguity in the definition of Delaunay triangulation if four or more Voronoi regions meet at a common point u. One such case is shown in Figure 1.3. The points generating the four or more regions all have the same distance from u: they lie on a common circle around u. Probabilistically, the chance of picking even just four points on a circle is zero because the circle defined by the first three points has zero measure in $\mathbb{R}^2$. A common way to say the same thing is that four points on a common circle form a *degeneracy* or a *special case*. An arbitrarily small perturbation suffices to remove the degeneracy and to reduce the special case to the general case.

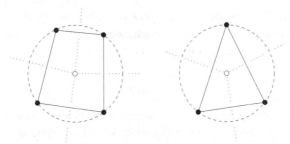

Figure 1.3. To the left, four dotted Voronoi edges meet at a common vertex and the dual Delaunay edges bound a quadrilateral. To the right, we have the general case, where only three Voronoi edges meet at a common vertex and the Delaunay edges bound a triangle.

We will often assume *general position*, which is the absence of any degeneracy. This really means that we delay the treatment of degenerate cases to later. The treatment is eventually done by perturbation, which can be actual or conceptual, or by exhaustive case analysis.

Circles and power

For now we assume general position. For a Delaunay triangle, *abc*, consider the circumcircle, which is the unique circle passing through *a*, *b*, and *c*. Its center is the corresponding Voronoi vertex, $u = V_a \cap V_b \cap V_c$, and its radius is $\varrho = \|u - a\| = \|u - b\| = \|u - c\|$; see Figure 1.3. We call the circle *empty* because it encloses no point of S. It turns out that empty circles characterize Delaunay triangles.

Circumcircle Claim. Let $S \subset \mathbb{R}^2$ be finite and in general position, and let $a, b, c \in S$ be three points. Then *abc* is a Delaunay triangle if and only if the circumcircle of *abc* is empty.

It is not entirely straightforward to see that this is true, at least not at the moment. Instead of proving the Circumcircle Claim, we focus our attention on a new concept of distance from a circle. The *power* of a point $x \in \mathbb{R}^2$ from a circle U with center u and radius ϱ is

$$\pi_U(x) = \|x - u\|^2 - \varrho^2.$$

If x lies outside the circle, then $\pi_U(x)$ is the square length of a tangent line segment connecting x with U. In any case, the power is positive outside the circle, zero on the circle, and negative inside the circle. We sometimes think of a circle as a weighted point and of the power as a weighted distance to that point. Given two circles, the set of points with equal power from both is a line. Figure 1.4 illustrates three different arrangements of two circles and their bisectors of points with equal power from both.

Acyclicity

We use the notion of power to prove an acyclicity result for Delaunay triangles. Let $x \in \mathbb{R}^2$ be an arbitrary but fixed viewpoint. We say a triangle *abc lies in front of* another triangle *def* if there is a half-line starting at x that first passes through *abc* and then through *def*; see Figure 1.6. We write $abc \prec def$ if *abc* lies in front of *def*. The set of Delaunay triangles together with $\prec$ forms a relation.

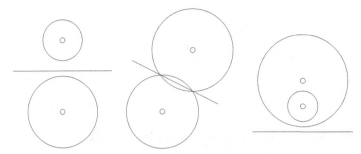

Figure 1.4. Three times two circles with bisector. From left to right: two disjoint and nonnested circles; two intersecting circles; and two nested circles.

General relations have cycles, which are sequences $\tau_0 \prec \tau_1 \prec \cdots \prec \tau_k \prec \tau_0$. Such cycles can also occur in general triangulations, as illustrated in Figure 1.5, but they cannot occur if the triangles are defined by empty circumcircles.

Acyclicity Lemma. The in-front relation for the set of Delaunay triangles defined by a finite set $S \subseteq \mathbb{R}^2$ is acyclic.

Proof. We show that $abc \prec def$ implies that the power of x from the circumcircle of abc is less than the power from the circumcircle of def. Define $abc = \tau_0$ and write $\pi_0(x)$ for the power of x from the circumcircle of abc. Similarly define $def = \tau_k$ and $\pi_k(x)$. Because S is finite, we can choose a half-line that starts at x, passes through abc and def, and contains no point of S. It intersects a sequence of Delaunay triangles:

$$abc = \tau_0 \prec \tau_1 \prec \cdots \prec \tau_k = def.$$

For any two consecutive triangles, the bisector of the two circumcircles contains the common edge. Because the third point of τ_{i+1} lies outside the circumcircle of τ_i, we have $\pi_i(x) < \pi_{i+1}(x)$ for $0 \leq i \leq k - 1$. Hence $\pi_0(x) < \pi_k(x)$. The

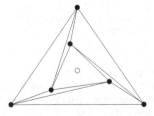

Figure 1.5. From the viewpoint in the middle, the three skinny triangles form a cycle in the in-front relation.

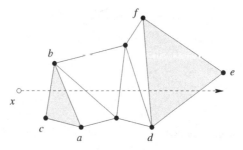

Figure 1.6. Triangle abc lies in front of triangle def. If abc and def belong to a Delaunay triangulation, then there is a sequence of triangles between them that all intersect the half-line.

acyclicity of the relation follows because real numbers cannot increase along a cycle.   □

Bibliographic notes

Voronoi diagrams are named after the Ukrainian mathematician Georges Voronoi, who published two seminal papers at the beginning of the twentieth century [5]. The same concept was discussed about half a century earlier by P. G. L. Dirichlet, and there are unpublished notes by René Descartes suggesting that he was using Voronoi diagrams in the first half of the seventeenth century. Delaunay triangulations are named after the Russian mathematician Boris Delaunay (also Delone), who dedicated his paper on empty spheres [2] to Georges Voronoi. The article by Franz Aurenhammer [1] offers a nice survey of Voronoi diagrams and their algorithmic applications. The acyclicity of Delaunay triangulations in arbitrary dimensions was proved by Edelsbrunner [3] and subsequently applied in computer graphics. In particular, the three-dimensional case has been exploited for the visualization of diffuse volumes [4, 6].

[1] F. Aurenhammer. Voronoi diagrams – a study of a fundamental geometric data structure. *ACM Comput. Surveys* **23** (1991), 345–405.

[2] B. Delaunay. Sur la sphère vide. *Izv. Akad. Nauk SSSR, Otdelenie Matematicheskii i Estestvennyka Nauk* **7** (1934), 793–800.

[3] H. Edelsbrunner. An acyclicity theorem for cell complexes in d dimensions. *Combinatorica* **10** (1990), 251–260.

[4] N. Max, P. Hanrahan, and R. Crawfis. Area and volume coherence for efficient visualization of 3D scalar functions. *Comput. Graphics* **24** (1990), 27–33.

[5] G. Voronoi. Nouvelles applications des paramètres continus à la théorie des formes quadratiques. *J. Reine Angew. Math.* **133** (1907), 97–178, and **134** (1908), 198–287.

[6] P. L. Williams. Visibility ordering meshed polyhedra. *ACM Trans. Graphics* **11** (1992), 103–126.

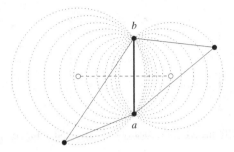

Figure 1.7. The Voronoi edge is the dashed line segment of centers of circles passing through the endpoints of *ab*.

1.2 Edge flipping

This section introduces a local condition for edges, shows it implies that a triangulation is Delaunay, and derives an algorithm based on edge flipping. The correctness of the algorithm implies that, among all triangulations of a given point set, the Delaunay triangulation maximizes the smallest angle.

Empty circles

Recall the Circumcircle Claim, which says that three points $a, b, c \in S$ are vertices of a Delaunay triangle if and only if the circle that passes through a, b, c is empty. A Delaunay edge, ab, belongs to one or two Delaunay triangles. In either case, there is a pencil of empty circles passing through a and b. The centers of these circles are the points on the Voronoi edge $V_a \cap V_b$; see Figure 1.7. What the Circumcircle Claim is for triangles, the Supporting Circle Claim is for edges.

Supporting Circle Claim. Let $S \subseteq \mathbb{R}^2$ be finite and in general position and let $a, b \in S$. Then ab is a Delaunay edge if and only if there is an empty circle that passes through a and b.

Delaunay lemma

By a *triangulation* we mean a collection of triangles together with their edges and vertices. A triangulation K *triangulates* S if the triangles decompose the convex hull of S and the set of vertices is S. An edge $ab \in K$ is *locally Delaunay* if

(i) it belongs to only one triangle and therefore bounds the convex hull, or
(ii) it belongs to two triangles, abc and abd, and d lies outside the circumcircle of abc.

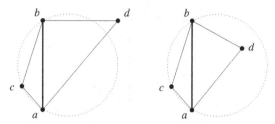

Figure 1.8. To the left *ab* is locally Delaunay and to the right it is not.

The definition is illustrated in Figure 1.8. A locally Delaunay edge is not necessarily an edge of the Delaunay triangulation, and it is fairly easy to construct
such an example. However, if *every* edge is locally Delaunay, then we can show
that *all* are Delaunay edges.

Delaunay Lemma. If every edge of K is locally Delaunay, then K is the Delaunay triangulation of S.

Proof. Consider a triangle $abc \in K$ and a vertex $p \in K$ different from a, b, c.
We show that p lies outside the circumcircle of abc. Because this is then true
for every p, the circumcircle of abc is empty, and because this is then true
for every triangle abc, K is the Delaunay triangulation of S. Choose a point x
inside abc such that the line segment from x to p contains no vertex other
than p. Let $abc = \tau_0, \tau_1, \ldots, \tau_k$ be the sequence of triangles that intersect
xp, as in Figure 1.9. We write $\pi_i(p)$ for the power of p to the circumcircle
of τ_i, as before. Since the edges along xp are all locally Delaunay, we have
$\pi_0(p) > \pi_1(p) > \cdots > \pi_k(p)$. Since p is one of the vertices of the last triangle,
we have $\pi_k(p) = 0$. Therefore $\pi_0(p) > 0$, which is equivalent to p's lying
outside the circumcircle of abc.   □

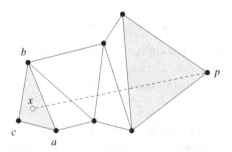

Figure 1.9. Sequence of triangles in K that intersect xp.

Edge-flip algorithm

If *ab* belongs to two triangles, *abc* and *abd*, whose union is a convex quadrangle, then we can *flip ab* to *cd*. Formally, this means we remove *ab*, *abc*, *abd* from the triangulation and we add *cd*, *acd*, *bcd* to the triangulation, as in Figure 1.10. The picture of a flip looks like a tetrahedron with the front and back superimposed. We can use edge flips as elementary operations to convert an arbitrary triangulation *K* to the Delaunay triangulation. The algorithm uses a stack and maintains the invariant that unless an edge is locally Delaunay, it resides on the stack. To avoid duplicates, we mark edges stored on the stack. Initially, all edges are marked and pushed on the stack.

```
while stack is non-empty do
    pop ab from stack and unmark it;
    if ab not locally Delaunay then
        flip ab to cd;
        for xy ∈ {ac, cb, bd, da} do
            if xy not marked then
                mark xy and push it on stack
            endif
        endfor
    endif
endwhile.
```

Let n be the number of points. The amount of memory used by the algorithm is $O(n)$ because there are at most $3n - 6$ edges, and the stack contains at most one copy of each edge. At the time the algorithm terminates, every edge is locally Delaunay. By the Delaunay Lemma, the triangulation is therefore the Delaunay triangulation of the point set.

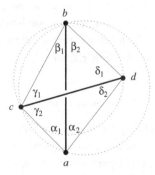

Figure 1.10. Flipping *ab* to *cd*. If *ab* is not locally Delaunay, then the union of the two triangles is convex and *cd* is locally Delaunay.

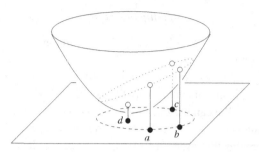

Figure 1.11. Points a, b, c lie on the dashed circle in the x_1x_2-plane and d lies inside that circle. The dotted curve is the intersection of the paraboloid with the plane that passes through $\hat{a}, \hat{b}, \hat{c}$. It is an ellipse whose projection is the dashed circle.

Circle and plane

Before proving the algorithm terminates, we interpret a flip as a tetrahedron in three-dimensional space. Let $\hat{a}, \hat{b}, \hat{c}, \hat{d}$ be the vertical projections of points a, b, c, d in the x_1x_2-plane onto the paraboloid defined as the graph of Π: $x_3 = x_1^2 + x_2^2$; see Figure 1.11.

Lifted Circle Claim. Point d lies inside the circumcircle of abc if and only if point $\hat{d}$ lies vertically below the plane passing through $\hat{a}, \hat{b}, \hat{c}$.

Proof. Let U be the circumcircle of abc and H the plane passing through $\hat{a}, \hat{b}, \hat{c}$. We first show that U is the vertical projection of $H \cap \operatorname{gf} \Pi$. Transform the entire space by mapping every point (x_1, x_2, x_3) to $(x_1, x_2, x_3 - x_1^2 - x_2^2)$. Points $\hat{a}, \hat{b}, \hat{c}, \hat{d}$ are mapped back to a, b, c, d, and the paraboloid Π becomes the x_1x_2-plane. The plane H becomes a paraboloid that passes through a, b, c. It intersects the x_1x_2-plane in the circumcircle of abc. Plane H partitions $\operatorname{gf} \Pi$ into a patch below H, a curve in H, and a patch above H. The curve in H is projected onto the circumcircle of abc, and the patch below H is projected onto the open disk inside the circle. It follows that $\hat{d}$ belongs to the patch below H if and only if d lies inside the circumcircle of abc. □

Running time

Flipping ab to cd is like gluing the tetrahedron $\hat{a}\hat{b}\hat{c}\hat{d}$ from below to $\hat{a}\hat{b}\hat{c}$ and $\hat{a}\hat{b}\hat{d}$. The algorithm can be understood as gluing a sequence of tetrahedra. Once we glue $\hat{a}\hat{b}\hat{c}\hat{d}$, we cannot glue another tetrahedron right below $\hat{a}\hat{b}$. In other words, once we flip ab we cannot introduce ab again by some other flip. This implies there are at most as many flips as there are edges connecting

Figure 1.12. To the left we see about one third of the edges in the initial triangulation, and to the right we see the same number of edges in the final Delaunay triangulation.

n points, namely $\binom{n}{2}$. Each flip takes constant time; hence the total running time is $O(n^2)$.

There are cases in which the algorithm takes $\Theta(n^2)$ flips to change an initial triangulation to the Delaunay triangulation, and one such case is illustrated in Figure 1.12. Take a convex upper and a concave lower curve and place m points on each, such that the upper points lie to the left of the lower points. The edges connecting the two curves in the initial and the Delaunay triangulation are shown in Figure 1.12. For each point, count the positions it is away from the middle, and for each edge charge the minimum of the two numbers obtained for its endpoints. In the initial triangulation, the total charge is about m^2, and in the Delaunay triangulation, the total charge is zero. Each flip moves an endpoint by at most one position and therefore decreases the charge by at most one. A lower bound of about m^2 for the number of flips follows.

MaxMin angle property

A flip substitutes two new triangles for two old triangles. It therefore changes six of the angles. In Figure 1.10, the new angles are $\gamma_1, \delta_1, \beta_1 + \beta_2, \gamma_2, \delta_2$, and $\alpha_1 + \alpha_2$, and the old angles are $\alpha_1, \beta_1, \gamma_1 + \gamma_2, \alpha_2, \beta_2$, and $\delta_1 + \delta_2$. We claim that for each of the six new angles there is an old angle that is at least as small. Indeed, $\gamma_1 \geq \alpha_2$ because both angles are opposite the same edge, namely bd, and a lies outside the circle passing through b, c, d. Similarly, $\delta_1 \geq \alpha_1, \gamma_2 \geq \beta_2$, and $\delta_2 \geq \beta_1$, and for trivial reasons $\beta_1 + \beta_2 \geq \beta_1$ and $\alpha_1 + \alpha_2 \geq \alpha_1$. It follows that a flip does not decrease the smallest angle in a triangulation. Since we can go from any triangulation K of S to the Delaunay triangulation, this implies that the smallest angle in K is no larger than the smallest angle in the Delaunay triangulation.

MaxMin Angle Lemma. Among all triangulations of a finite set $S \subseteq \mathbb{R}^2$, the Delaunay triangulation maximizes the minimum angle.

Figure 1.13 illustrates the above proof of the MaxMin Angle Lemma by sketching what we call the *flip graph* of S. Each triangulation is a node, and there is a

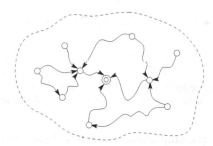

Figure 1.13. Sketch of flip graph. The sink is the Delaunay triangulation. There is a directed path from every node to the Delaunay triangulation.

directed arc from node μ to node ν if there is a flip that changes the triangulation μ to ν. The direction of the arc corresponds to our requirement that the flip substitutes a locally Delaunay edge for one that is not locally Delaunay. The running time analysis implies that the flip graph is acyclic and that its undirected version is connected. If we allow flips in either direction, we can go from any triangulation of S to any other triangulation in less than n^2 flips.

Bibliographic notes

A proof of the Delaunay Lemma and its generalization to arbitrary finite dimensions is contained in the original paper by Boris Delaunay [1]. The edge-flip algorithm is from Charles Lawson [3]. The algorithm does not generalize to three or higher dimensions. For planar triangulations, the edge-flip operation is widely used to improve local quality measures; see, for example, Schumaker [4]. Unfortunately, the algorithms get caught in local optima for almost all interesting measures. The observation that the Delaunay triangulation maximizes the smallest angle was first made by Robin Sibson [5]. Minimizing the largest angle seems more difficult, and the only known polynomial time algorithm uses edge insertions, which are somewhat more powerful than edge flips [2].

[1] B. Delaunay. Sur la sphère vide. *Izv. Akad. Nauk SSSR, Otdelenie Matematicheskii i Estestvennyka Nauk* **7** (1934), 793–800.
[2] H. Edelsbrunner, T. S. Tan, and R. Waupotitsch. An O($n^2 \log n$) time algorithm for the minmax angle triangulation. *SIAM J. Sci. Stat. Comput.* **13** (1992), 994–1008.
[3] C. L. Lawson. Software for C^1 surface interpolation. In *Mathematical Software III*, Academic Press, New York, 1977, 161–194.
[4] L. L. Schumaker. Triangulation methods. In *Topics in Multivariate Approximation*, C. K. Choi, L. L. Schumaker, and F. I. Utreras (eds.), Academic Press, New York, 1987, 219–232.
[5] R. Sibson. Locally equiangular triangulations. *Comput. J.* **21** (1978), 243–245.

1.3 Randomized construction

The algorithm in this section constructs Delaunay triangulations incrementally, using edge flips and randomization. After explaining the algorithm, we present a detailed analysis of the expected amount of resources it requires.

Incremental algorithm

We obtain a fast algorithm for constructing Delaunay triangulations if we interleave flipping edges with adding points. Denote the points in $S \subseteq \mathbb{R}^2$ as $p_1, p_2, \ldots, p_n$ and assume general position. When we add a point to the triangulation, it can either lie inside or outside the convex hull of the preceding points. To reduce the outside case to the inside case, we start with a triangulation D_0 that consists of a single and sufficiently large triangle xyz. Define $S_i = \{x, y, z, p_1, p_2, \ldots, p_i\}$, and let D_i be the Delaunay triangulation of S_i. The algorithm is a for-loop adding the points in sequence. After adding a point, it uses edge flips to satisfy the Delaunay Lemma before the next point is added.

```
for i = 1 to n do
    find τᵢ₋₁ ∈ Dᵢ₋₁ containing pᵢ;
    add pᵢ by splitting τᵢ₋₁ into three;
    while ∃ab not locally Delaunay do
        flip ab to other diagonal cd
    endwhile
endfor.
```

The two elementary operations used by the algorithm are shown in Figure 1.14. Both pictures can be interpreted as the projection of a tetrahedron, though from different angles. For this reason, the addition of a point inside a triangle is sometimes called a 1-to-3 flip, while an edge flip is sometimes also called a 2-to-2 flip.

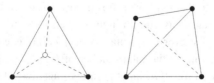

Figure 1.14. To the left, the hollow vertex splits the triangle into three. To the right, the dashed diagonal replaces the solid diagonal.

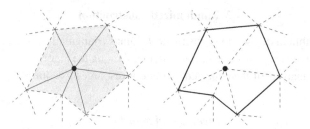

Figure 1.15. The star of the solid vertex is shown on the left and the link of the same vertex is shown on the right.

Growing star

Note that every new triangle in D_i has p_i as one of its vertices. Indeed, *abc* is a triangle in D_i if and only if $a, b, c \in S_i$ and the circumcircle is empty of points in S_i. But if p_i is not one of the vertices, then $a, b, c \in S_{i-1}$, and if the circumcircle is empty of points in S_i, then it is also empty of points in S_{i-1}. So *abc* is also a triangle in D_{i-1}. This implies that all flips during the insertion of p_i occur right around p_i.

We need some definitions. The *star* of p_i consists of all triangles that contain p_i. The *link* of p_i consists of all edges of triangles in the star that are disjoint from p_i. Both concepts are illustrated in Figure 1.15. Right after p_i is added, the link consists of three edges, namely the edges of the triangle that contains p_i. These edges are marked and pushed on the stack to start the edge-flipping while-loop. Each flip replaces a link edge by an edge with endpoint p_i. At the same time, it removes one triangle in the star and one outside the star and it adds the two triangles that cover the same quadrangle to the star. The net effect is one more triangle in the star. The number of edge flips is therefore three less than the number of edges in the final link, which is the same as three less than the degree of p_i in D_i.

Number of flips

We temporarily ignore the time needed to find the triangles τ_{i-1}. The rest of the time is proportional to the number of flips needed to add p_1 to p_n. We assume $p_1, p_2, \ldots, p_n$ is a randomly chosen input sequence. Random does not mean arbitrary but rather that every permutation of n points is equally likely. The expected number of flips is the total number of flips needed to construct the Delaunay triangulation for all $n!$ input permutations divided by $n!$.

Consider inserting the last point, p_n. The sum of degrees of all possible last points is the same as the sum of degrees of all points p_i in D_n. The latter is equal to twice the number of edges and therefore

$$\sum_{i=1}^{n} \deg p_i \le 6n.$$

The number of flips needed to add all last points is therefore at most $6n - 3n = 3n$. Each last point is added $(n - 1)!$ times. The total number of flips is therefore

$$F(n) \le n \cdot F(n - 1) + 3n!$$

$$\le 3n \cdot n!.$$

Indeed, if we assume $F(n - 1) \le 3(n - 1) \cdot (n - 1)!$, we get $n \cdot F(n - 1) + 3n! \le 3(n - 1) \cdot n! + 3n! = 3n \cdot n!$. The expected number of edge flips needed for n points is therefore at most $3n$.

There is a simple way to say the same thing. The expected number of flips for the last point is at most three, and therefore the expected number of flips to add any point is at most three.

The history

We use the evolution of the Delaunay triangulation to find the triangle τ_{i-1} that contains point p_i. Instead of deleting a triangle when it is split or flipped away, we just make it the parent of the new triangles. Figure 1.16 shows the two operations to the left and the corresponding parent–child relations to the right. Each time we split or flip, we add triangles or nodes to the growing data

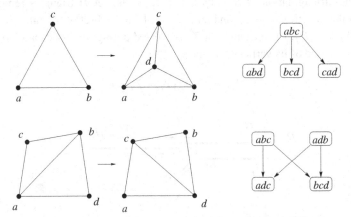

Figure 1.16. Splitting a triangle generates a parent with three children. Flipping an edge generates two parents sharing the same two children.

structure that records the history of the construction. The evolution from D_0 to D_n consists of n splits and an expected number of at most $3n$ flips. The resulting directed acyclic graph, or DAG for short, therefore has an expected size of at most $1 + 3n + 2 \cdot 3n = 9n + 1$ nodes. It has a unique source, the triangle xyz, and its sinks are the triangles in D_n.

Searching and charging

Consider adding the point p_i. To find the triangle $\tau_{i-1} \in D_{i-1}$, we search a path of triangles in the history DAG that all contain p_i. The path begins as xyz and ends at τ_{i-1}. The history DAG of D_{i-1} consists of i layers. Layers 0 to j represent the DAG of D_j. Its sinks are the triangles in D_j, and we let $\sigma_j \in D_j$ be the triangle that contains p_i. Triangles σ_0 to σ_j form a not necessarily contiguous subsequence of nodes along the search path. It is quite possible that some of the triangles σ are the same. Let G_j be the set of triangles removed from D_j during the insertion of p_{j+1}, and let H_j be the set of triangles removed from D_j during the hypothetical and independent insertion of p_i into D_j. The two sets are schematically sketched as intervals along the real line representing the Delaunay triangulation in Figure 1.17. We have $\sigma_j = \sigma_{j+1}$ if G_j and H_j are disjoint. Suppose $\sigma_j \neq \sigma_{j+1}$. Then $X_j = G_j \cap H_j \neq \emptyset$, and all triangles on the portion of the path from σ_j to σ_{j+1} are generated by flips that remove triangles in X_j. The cost for searching with p_i is therefore at most proportional to the sum of card X_j, for j from 0 to $i - 2$.

We write X_j in terms of other sets. These sets represent what happens if we again hypothetically first insert p_i into D_j and then insert p_{j+1} into the Delaunay triangulation of $S_j \cup \{p_i\}$. Let Y_j be the set of triangles removed during the insertion of p_{j+1}, and let $Z_j \subseteq Y_j$ be the subset of triangles that do not belong to D_j. Each triangle in Z_j is created during the insertion of p_i, so p_i must be one of its vertices. We have

$$X_j = G_j - (Y_j - Z_j).$$

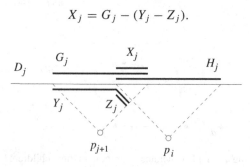

Figure 1.17. The intervals represent sets of triangles removed or added when we insert p_{j+1} and/or p_i to D_j.

Expectations

We bound the expected search time by bounding the expected total size of the X_j. Write cardinalities by using corresponding lower-case letters. Because $Z_j \subseteq Y_j$ and $Y_j - Z_j \subseteq G_j$, we have

$$x_j = g_j - y_j + z_j.$$

The expected values of g_j and y_{j-1} are the same, because both count triangles removed by inserting a random jth point. Because the expectation of a sum is the sum of expectations, we have

$$E\left[\sum_{j=0}^{i-2} x_j\right] = \sum_{j=0}^{i-2} E[g_j] - E[y_j] + E[z_j]$$

$$= E[g_0 - g_{i-1}] + \sum_{j=0}^{i-2} E[z_j].$$

To compute the expected value of z_j, we use the fact that among $j + 2$ points, every pair is equally likely to be p_{j+1} and p_i. For example, if p_{j+1} and p_i are not connected by an edge in the Delaunay triangulation of $S_j \cup \{p_{j+1}, p_i\}$, then $Z_j = \emptyset$. In general, a triangle in the Delaunay triangulation of $S_j \cup \{p_i\}$ has probability at most $3/(j + 1)$ of being in the star of p_i. The expected number of triangles removed by inserting p_{j+1} is at most four. Because the expectation of a product is the product of expectations, we have $E[z_j] \leq (4 \cdot 3)/(j + 1)$. The expected length of the search path for p_i is

$$\sum_{j=0}^{i-2} E[x_j] \leq \sum_{j=0}^{i-2} \frac{12}{j + 1} \leq 1 + 12 \ln(i - 1).$$

The expected total time spent on searching in the history DAG is $\sum_i \sum_j E[x_j] \leq c \cdot n \log n$.

To summarize, the randomized incremental algorithm constructs the Delaunay triangulation of n points in $\mathbb{R}^2$ in the expected time $O(n \log n)$ and with the expected amount of memory $O(n)$.

Bibliographic notes

The randomized incremental algorithm of this section is from Guibas, Knuth, and Sharir [3]. It has been generalized to three and higher dimensions by Edelsbrunner and Shah [2]. All this is based on earlier work on randomized algorithms and in particular on the methods developed by Clarkson and Shor [1]. The arguments used to bound the expected number of flips and the expected search time are examples of the backwards analysis introduced by Raimund Seidel [4].

[1] K. L. Clarkson and P. W. Shor. Applications of random sampling in computational geometry. *Discrete Comput. Geom.* **4** (1989), 387–421.
[2] H. Edelsbrunner and N. R. Shah. Incremental topological flipping works for regular triangulations. *Algorithmica* **15** (1996), 223–241.
[3] L. J. Guibas, D. E. Knuth, and M. Sharir. Randomized incremental construction of Delaunay and Voronoi diagrams. *Algorithmica* **7** (1992), 381–413.
[4] R. Seidel. Backwards analysis of randomized geometric algorithms. In *New Trends in Discrete and Computational Geometry*, J. Pach (ed.), Springer-Verlag, Berlin, 1993, 37–67.

1.4 Symbolic perturbation

The computational technique of symbolically perturbing a geometric input justifies the mathematically convenient assumption of general position. This section describes a particular perturbation known as SoS, or Simulation of Simplicity.

Orientation test

Let $a = (\alpha_1, \alpha_2)$, $b = (\beta_1, \beta_2)$, and $c = (\gamma_1, \gamma_2)$ be three points in the plane. We consider a, b, c degenerate if they lie on a common line. This includes the case in which two or all three points are the same. In the degenerate case, point c is an affine combination of a and b; that is, $c = \lambda_1 a + \lambda_2 b$ with $\lambda_1 + \lambda_2 = 1$. Such λ_1, λ_2 exist if and only if the determinant of

$$\Delta = \begin{bmatrix} 1 & \alpha_1 & \alpha_2 \\ 1 & \beta_1 & \beta_2 \\ 1 & \gamma_1 & \gamma_2 \end{bmatrix}$$

vanishes. In the nondegenerate case, the sequence a, b, c either forms a left or a right turn. We can again use the determinant of Δ to decide which it is.

Orientation Claim. The sequence a, b, c forms a left turn if and only if $\det \Delta > 0$, and it forms a right turn if and only if $\det \Delta < 0$.

Proof. We first check the claim for $a_0 = (0, 0)$, $b_0 = (1, 0)$, and $c_0 = (0, 1)$. It is geometrically obvious that a_0, b_0, c_0 form a left turn, and indeed

$$\det \begin{bmatrix} 1 & 0 & 0 \\ 1 & 1 & 0 \\ 1 & 0 & 1 \end{bmatrix} = 1.$$

We can continuously move a_0, b_0, c_0 to any other left-turn a, b, c without ever having three collinear points. Since the determinant changes continuously with the coordinates, it remains positive during the entire motion and is therefore positive at a, b, c. Symmetry implies that all right turns have negative determinants.   □

In-circle test

The in-circle test is formulated for four points a, b, c, d in the plane. We consider a, b, c, d degenerate if a, b, c lie on a common line or a, b, c, d lie on a common circle. We already know how to test for points on a common line. To test for points on a common circle, we recall the definition of lifted points, $\hat{a} = (\alpha_1, \alpha_2, \alpha_3)$ with $\alpha_3 = \alpha_1^2 + \alpha_2^2$, and so on. Points a, b, c, d lie on a common circle if and only if $\hat{a}, \hat{b}, \hat{c}, \hat{d}$ lie on a common plane in $\mathbb{R}^3$; see Figure 1.11. In other words, $\hat{d}$ is an affine combination of $\hat{a}, \hat{b}, \hat{c}$, which is equivalent to

$$
\Gamma = \begin{bmatrix} 1 & \alpha_1 & \alpha_2 & \alpha_3 \\ 1 & \beta_1 & \beta_2 & \beta_3 \\ 1 & \gamma_1 & \gamma_2 & \gamma_3 \\ 1 & \delta_1 & \delta_2 & \delta_3 \end{bmatrix}
$$

having zero determinant. In the nondegenerate case, d either lies inside or outside the circle defined by a, b, c. We can use the determinants of Δ and Γ to decide which it is. Note that permuting a, b, c can change the sign of det Γ without changing the geometric configuration. Since the signs of det Γ and det Δ change simultaneously, we can counteract by multiplying the two.

In-circle Claim. Point d lies inside the circle passing through a, b, c if and only if det $\Delta \cdot$ det $\Gamma < 0$, and d lies outside the circle if and only if det $\Delta \cdot$ det $\Gamma > 0$.

Proof. We first check the claim for $d_0 = (^1\!/_2, {}^1\!/_2)$ and $a_0 = (0, 0)$, $b_0 = (1, 0)$, and $c_0 = (0, 1)$ as before. Point d_0 lies at the center and therefore inside the circle passing through a_0, b_0, c_0. The determinant of Δ is 1, and that of Γ is

$$
\det \begin{bmatrix} 1 & 0 & 0 & 0 \\ 1 & 1 & 0 & 1 \\ 1 & 0 & 1 & 1 \\ 1 & \frac{1}{2} & \frac{1}{2} & \frac{1}{2} \end{bmatrix} = -\frac{1}{2},
$$

so their product is negative. As in the proof of the Orientation Claim, we derive the general result from the special one by continuity. Specifically, every configuration a, b, c, d, where d lies inside the circle of a, b, c, can be obtained from a_0, b_0, c_0, d_0 by continuous motion, avoiding all degeneracies. The signs of the two determinants remain the same throughout the motion, and so does their product. This implies the claim for negative products, and symmetry implies the claim for positive products. $\qquad\square$

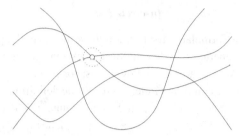

Figure 1.18. Schematic picture of the union of $(2n - 1)$-dimensional manifolds in $2n$-dimensional space. The marked point lies on two manifolds and thus has two degenerate subconfigurations. The dotted circle bounds a neighbourhood, and most points in that neighbourhood are non-degenerate.

Algebraic framework

Let us now take a more abstract and algebraic view of degeneracy as a geometric phenomenon. For expository reasons, we restrict ourselves to orientation tests in the plane. Let S be a collection of n points, denoted as $p_i = (\phi_{i,1}, \phi_{i,2})$, for $1 \leq i \leq n$. By listing the $2n$ coordinates in a single sequence, we think of S as a single point in $2n$-dimensional space. Specifically, S is mapped to $Z = (\zeta_1, \zeta_2, \zeta_3, \ldots, \zeta_{2n}) \in \mathbb{R}^{2n}$, where $\zeta_{2i-1} = \phi_{i,1}$ and $\zeta_{2i} = \phi_{i,2}$, for $1 \leq i \leq n$. Point Z is degenerate if and only if

$$
\det \begin{bmatrix} 1 & \zeta_{2i-1} & \zeta_{2i} \\ 1 & \zeta_{2j-1} & \zeta_{2j} \\ 1 & \zeta_{2k-1} & \zeta_{2k} \end{bmatrix} = 0
$$

for some $1 \leq i < j < k \leq n$. The equation identifies a differentiable $(2n - 1)$-dimensional manifold in $\mathbb{R}^{2n}$. There are $\binom{n}{3}$ such manifolds, $\mathbb{M}_\ell$, and Z is degenerate if and only if $Z \in \bigcup_\ell \mathbb{M}_\ell$, as sketched in Figure 1.18. Each manifold has dimension one less than the ambient space and hence measure zero in $\mathbb{R}^{2n}$. We have a finite union of measure zero sets, which still has measure zero. In other words, most points in an open neighborhood of $Z \in \mathbb{R}^{2n}$ are non-degenerate. A point nearby Z is often called a perturbation of Z or S. The result on neighborhoods thus implies that there are arbitrarily close nondegenerate perturbations of S.

Perturbation

We construct a nondegenerate perturbation of S by using positive parameters $\varepsilon_1, \varepsilon_2, \ldots, \varepsilon_{2n}$. These parameters will be chosen anywhere between arbitrarily and sufficiently small, and we may think of them as infinitesimals. They will also be chosen sufficiently different, and we will see shortly what this means. Let $Z \in \mathbb{R}^{2n}$, and for every $\varepsilon > 0$ define

$$Z(\varepsilon) = (\zeta_1 + \varepsilon_1, \zeta_2 + \varepsilon_2, \ldots, \zeta_{2n} + \varepsilon_{2n}),$$

where $\varepsilon_i = f_i(\varepsilon)$ with $f_i : \mathbb{R} \to \mathbb{R}$ continuous and $f_i(0) = 0$. If the ε_i are sufficiently different, we get the following three properties provided $\varepsilon > 0$ is sufficiently small.

I. $Z(\varepsilon)$ is nondegenerate.
II. $Z(\varepsilon)$ retains all nondegenerate properties of Z.
III. The computational overhead for simulating $Z(\varepsilon)$ is negligible.

For example, if $\varepsilon_i = \varepsilon^{2^i}$ then $\varepsilon_1 \gg \varepsilon_2 \gg \cdots \gg \varepsilon_{2n}$ and we can do all computations simply by comparing indices without ever computing a feasible ε. We demonstrate this by explicitly computing the orientation of the points p_i, p_j, p_k after perturbation. By definition, that orientation is the sign of the determinant of

$$\Delta(\varepsilon) = \begin{bmatrix} 1 & \zeta_{2i-1} + \varepsilon_{2i-1} & \zeta_{2i} + \varepsilon_{2i} \\ 1 & \zeta_{2j-1} + \varepsilon_{2j-1} & \zeta_{2j} + \varepsilon_{2j} \\ 1 & \zeta_{2k-1} + \varepsilon_{2k-1} & \zeta_{2k} + \varepsilon_{2k} \end{bmatrix}.$$

Note that $\Delta(\varepsilon)$ is a polynomial in ε. The terms with smaller power are more significant than those with larger power. We assume $i < j < k$ and list the terms of $\Delta(\varepsilon)$ in the order of decreasing significance; that is,

$$\det \Delta(\varepsilon) = \det \Delta - \det \Delta_1 \cdot \varepsilon^{2^{2i-1}}$$
$$+ \det \Delta_2 \cdot \varepsilon^{2^{2i}} + \det \Delta_3 \cdot \varepsilon^{2^{2j-1}}$$
$$- 1 \cdot \varepsilon^{2^{2j-1}} \varepsilon^{2^{2i}} \pm \ldots,$$

where

$$\Delta = \begin{bmatrix} 1 & \zeta_{2i-1} & \zeta_{2i} \\ 1 & \zeta_{2j-1} & \zeta_{2j} \\ 1 & \zeta_{2k-1} & \zeta_{2k} \end{bmatrix},$$

$$\Delta_1 = \begin{bmatrix} 1 & \zeta_{2j} \\ 1 & \zeta_{2k} \end{bmatrix},$$

$$\Delta_2 = \begin{bmatrix} 1 & \zeta_{2j-1} \\ 1 & \zeta_{2k-1} \end{bmatrix},$$

$$\Delta_3 = \begin{bmatrix} 1 & \zeta_{2i} \\ 1 & \zeta_{2k} \end{bmatrix}.$$

Property I is satisfied because the fifth term is nonzero, and its influence on the sign of the determinant cannot be canceled by subsequent terms. Property II is satisfied because the sign of the perturbed determinant is the same as that of the unperturbed one, unless the latter vanishes.

Implementation

In order to show Property III, we give an implementation of the test for $Z(\varepsilon)$. First we sort the indices such that $i < j < k$, and we count the number of transpositions. Then we determine whether the three perturbed points form a left or a right turn by computing determinants of the four submatrices listed above.

```
boolean LEFTTURN(integer  i, j, k):
    assert i < j < k;
    case det Δ ≠ 0: return det Δ > 0;
    case det Δ₁ ≠ 0: return det Δ₁ < 0;
    case det Δ₂ ≠ 0: return det Δ₂ > 0;
    case det Δ₃ ≠ 0: return det Δ₃ > 0;
    otherwise: return FALSE.
```

If the number of transpositions needed to sort i, j, k is odd, then the sorting reverses the sign, and we correct the reversal by reversing the result of the Function LEFTTURN.

As an important detail, we note that signs of determinants have to be computed exactly. With normal floating point arithmetic, this is generally not possible. We must therefore resort to exact arithmetic methods using long integer or other representations of coordinates. These methods are typically more costly than floating point arithmetic, but differences vary widely among different computer hardware. A pragmatic compromise uses floating point arithmetic together with error analysis. After computing the determinant with floating point arithmetic, we check whether the absolute value is large enough for its sign to be guaranteed. Only if that guarantee cannot be obtained do we repeat the computation in exact arithmetic.

Bibliographic notes

The idea of using symbolic perturbation for computational reasons is already present in the work of George Danzig on linear programming [1]. It reappeared in computational geometry with the work of four independent groups of authors. Edelsbrunner and Mücke [2] develop SoS, which is the method described in this section. Yap [7] studies the class of perturbations obtained with different orderings of infinitesimals. Emiris and Canny [3] introduce perturbations along straight lines. Michelucci [5] exploits randomness in the design of perturbations.

Symbolic perturbations as a general computational technique within computational geometry remains a controversial subject. It succeeds in extending partially to completely correct software for some but not all geometric problems. Seidel [6] addresses this issue, offers a unified view of symbolic perturbation, and discusses limitations of the method. Fortune and Van Wyk [4] describe a floating point filter that reduces the overhead needed for exact computation.

[1] G. B. Danzig. *Linear Programming and Extensions*. Princeton Univ. Press, Princeton, New Jersey, 1963.
[2] H. Edelsbrunner and E. P. Mücke. Simulation of simplicity: a technique to cope with degenerate cases in geometric algorithms. *ACM Trans. Graphics* **9** (1990), 66–104.
[3] I. Emiris and J. Canny. A general approach to removing geometric degeneracies. *SIAM J. Comput.* **24** (1995), 650–664.
[4] S. Fortune and C. J. Van Wyk. Static analysis yields efficient exact integer arithmetic for computational geometry. *ACM Trans. Graphics* **15** (1996), 223–248.
[5] D. Michelucci. An ε-arithmetic for removing degeneracies. In "Proc. IEEE Sympos. Comput. Arithmetic," 1995.
[6] R. Seidel. The nature and meaning of perturbations in geometric computing. *Discrete Comput. Geom.* **19** (1998), 1–18.
[7] C. K. Yap. Symbolic treatment of geometric degeneracies. *J. Symbolic Comput.* **10** (1990), 349–370.

Exercise collection

The credit assignment reflects a subjective assessment of difficulty. A typical question can be answered by using knowledge of the material combined with some thought and analysis.

1. **Section of triangulation** (two credits). Let K be a triangulation of a set of n points in the plane. Let ℓ be a line that avoids all points. Prove that ℓ intersects at most $2n - 4$ edges of K and that this upper bound is tight for every $n \geq 3$.
2. **Minimum spanning tree** (one credit). The notion of a minimum spanning tree can be extended from weighted graphs to a geometric setting in which the nodes are points in the plane. Take the complete graph of the set of nodes and define the length of an edge as the Euclidean distance between its endpoints. A minimum spanning tree of that graph is a *Euclidean minimum spanning tree* of the point set. Prove that all edges of every Euclidean minimum spanning tree belong to the Delaunay triangulation of the same point set.

3. **Sorted angle vector** (one credit). Let K be a triangulation of a finite set in the plane. Let t be the number of triangles and consider the sorted vector of angles,

$$\mathbf{v}(K) = (\alpha_1 \le \alpha_2 \le \cdots \le \alpha_{3t}).$$

   Prove that $\mathbf{v}(K) = \mathbf{v}(D)$ or $\mathbf{v}(K)$ is lexicographically smaller than $\mathbf{v}(D)$, where D is the Delaunay triangulation of the points.

4. **Minmax circumcircle** (two credits). Let K be a triangulation of a finite set in the plane and let $\varrho(K)$ be the radius of the largest circumcircle of any triangle in K. Prove $\varrho(K) \ge \varrho(D)$, where D is the Delaunay triangulation of the set.

5. **Random permutation** (one credit). Show that the following algorithm constructs a random permutation of the integers 1 to n.

   ```
   for i = 1 to n do
       Z[i] = i; choose random index 1 ≤ j ≤ i;
       swap Z[i] and Z[j]
   endfor.
   ```

6. **Furthest-point Voronoi diagram** (one credit). Let $S \subseteq \mathbb{R}^2$ be finite. The *furthest-point Voronoi region* of a point $p \in S$ consists of all points at least as far from p as from any other point in S,

$$F_p = \{x \in \mathbb{R}^2 \mid \|x - p\| \ge \|x - q\|, \forall q \in S\}.$$

   (i) Prove $F_p \ne \emptyset$ if and only if p lies on the boundary of the convex hull of S.

   (ii) Draw the furthest-point Voronoi regions of about ten points in the plane, together with the dual furthest-point Delaunay triangulation.

7. **Line segment intersection** (two credits). Let a, b, x, y be points in $\mathbb{R}^2$. They are in general position if no three are collinear.

   (i) Assume general position and write a boolean function that decides whether the line segments ab and xy cross or are disjoint.

   (ii) What are the degenerate cases, and how does your function deal with them?

8. **Enumerating degeneracies** (one credit). Let a, b, c, d be points in $\mathbb{R}^3$. The *orientation* of the sequence is the sign of

$$\det \begin{bmatrix} 1 & \alpha_1 & \alpha_2 & \alpha_3 \\ 1 & \beta_1 & \beta_2 & \beta_3 \\ 1 & \gamma_1 & \gamma_2 & \gamma_3 \\ 1 & \delta_1 & \delta_2 & \delta_3 \end{bmatrix}.$$

Simulation of simplicity expands the determinant into a polynomial $P(\varepsilon)$, and the orientation is decided by finding the sign of P for sufficiently small $\varepsilon > 0$.

(i) List the terms of the polynomial in the order of decreasing significance.

(ii) The perturbation classifies and disambiguates the various degenerate cases that occur. Each class corresponds to a prefix of the polynomial that is identically zero. Describe each class in words or figures.

2

Triangle meshes

The three sections in this chapter apply what we learned in Chapter 1 to the construction of triangle meshes in the plane. In mesh generation, the vertices are no longer part of the input but have to be placed by the algorithm itself. A typical instance of the meshing problem is given as a region, and the algorithm is expected to decompose that region into cells or elements. This chapter focuses on constructing meshes with triangle elements, and it pays attention to quality criteria, such as angle size and length variation. Section 2.1 shows how Delaunay triangulations can be adapted to constraints given as line segments that are required to be part of the mesh. Section 2.2 and 2.3 describe and analyze the Delaunay refinement method that adds new vertices at circumcenters of already existing Delaunay triangles.

2.1 Constrained triangulations

This section studies triangulations in the plane constrained by edges specified as part of the input. We show that there is a unique constrained triangulation that is closest, in some sense, to the (unconstrained) Delaunay triangulation.

Constraining line segments

The preceding sections constructed triangulations for a given set of points. The input now consists of a finite set of points, $S \subseteq \mathbb{R}^2$, together with a finite set of line segments, L, each connecting two points in S. We require that any two line segments are either disjoint or meet at most in a common endpoint. A *constrained triangulation* of S and L is a triangulation of S that contains all line segments of L as edges. Figure 2.1 illustrates that we can construct a constrained triangulation by adding straight edges connecting points in S as long as they have no interior points in common with previous edges.

26

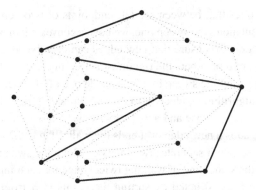

Figure 2.1. Given the points and solid edges, we form a constrained triangulation by adding as many dotted edges as possible without creating improper intersections.

Plane-sweep algorithm

The idea of organizing the actions of the algorithm around a line sweeping over the plane leads to an efficient way of constructing constrained triangulations. We use a vertical line that sweeps over the plane from left to right, as shown in Figure 2.2. The algorithm uses two data structures. The *schedule*, X, orders events in time. The *cross-section*, Y, stores the line segments in L that currently intersect the sweep line. The algorithm is defined by the following invariant.

Invariant (I) At any moment in time, the partial triangulation contains all edges in L, a maximal number of edges connecting points to the left of the sweep line, and no other edges.

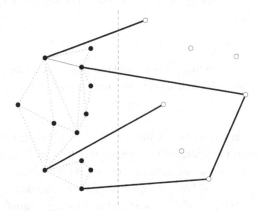

Figure 2.2. Snapshot of plane-sweep construction of a constrained triangulation.

Invariant (I) implies that between the left endpoints of two constraining line
segments adjacent along the sweep line, we have a convex chain of edges in the
partial triangulation. To ensure that new edges can each be added in constant
time, the algorithm remembers the rightmost vertex in each chain. If the point
p encountered next by the sweep line falls inside one of the intervals along the
sweep line, the algorithm connects p to the corresponding rightmost vertex. It
then proceeds in a clockwise and anticlockwise order along the convex chain.
Each step either adds a new edge or it ends the walk. If p is the right endpoint
of a line segment, then it separates two intervals along the sweep line, and the
algorithm does the same kind of walking twice, once for each interval.

The schedule is constructed by sorting the points in S from left to right,
which can be done in time $O(n \log n)$, where $n = \operatorname{card} S$. The cross-section is
maintained as a dictionary, which supports search, insertion, and deletion all in
time $O(\log n)$. There is a search for each point in S and an insertion–deletion pair
for each line segment in L, taking total time $O(n \log n)$. Fewer than $3n$ edges are
added to the triangulation, each in constant time. The plane-sweep algorithm
thus constructs a constrained triangulation of S and L in time $O(n \log n)$.

Constrained Delaunay triangulations

The triangulations constructed by a plane sweep usually have many small and
large angles. We use a notion of visibility between points to introduce a constrained triangulation that avoids small angles to the extent possible.

Points $x, y \in \mathbb{R}^2$ are *visible* from each other if xy contains no point of S in its
interior and it shares no interior point with a constraining line segment. Formally,
$\operatorname{int} xy \cap S = \emptyset$ and $\operatorname{int} xy \cap uv = \emptyset$ for all $uv \in L$. Assume general position. An
edge ab, with $a, b \in S$, belongs to the *constrained Delaunay triangulation* of S
and L if

(i) $ab \in L$, or
(ii) a and b are visible from each other and there is a circle passing through a
      and b such that each point inside this circle is invisible from every point
      $x \in \operatorname{int} ab$.

We say the circle in (ii) *witnesses* the membership of ab in the constrained
Delaunay triangulation. Figure 2.3 illustrates this definition. Note if $L = \emptyset$ then
the constrained Delaunay triangulation of S and L is the Delaunay triangulation
of S. More generally, it is, however, unclear that what we defined is indeed a
triangulation. For example, why is it true that no two edges satisfying (i) or (ii)
cross?

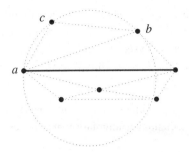

Figure 2.3. Constrained Delaunay triangulation for seven points and one constraining line segment. The circumcircle of *abc* encloses only points that are invisible from all points of int *ab*.

Edge flipping

We introduce a generalized concept of being locally Delaunay, and we use it to prove that the above definition makes sense. Let K be any constrained triangulation of S and L. An edge $ab \in K$ is *locally Delaunay* if $ab \in L$, or ab is a convex hull edge, or d lies outside the circumcircle of abc, where $abc, abd \in K$.

Constrained Delaunay Lemma. If every edge of K is locally Delaunay, then K is the constrained Delaunay triangulation of S and L.

Proof. We show that every edge in K satisfies (i) or (ii) and therefore belongs to the constrained Delaunay triangulation. The claim follows because every additional edge crosses at least one edge of K and therefore of the constrained Delaunay triangulation.

Let ab be an edge and p a vertex in K. Assume $ab \notin L$, for else ab belongs to the constrained Delaunay triangulation for trivial reasons. Assume also that ab is not a convex hull edge, for else we can easily find a circle passing through a and b such that p lies outside the circle. Hence, ab belongs to two triangles, and we let abc be the one separated from p by the line passing through ab. We need to prove that if p is visible from a point $x \in \text{int } ab$, then it lies outside the circumcircle of abc. Consider the sequence of edges in K crossing xp. Since x and p are visible from each other, all these edges are not in L. We can therefore apply the argument of the proof of the original Delaunay Lemma, which is illustrated in Figure 1.9. □

This result suggests we use the edge-flipping algorithm to construct the constrained Delaunay triangulation. The only difference to the original edge-flipping algorithm is that edges in L are not flipped, since they are locally Delaunay by definition. As before, the algorithm halts in time $\Omega(n^2)$ after fewer than $\binom{n}{2}$ flips. The analysis of angle changes during an edge flip presented in Section 1.2 implies that the MaxMin Angle Lemma also holds in the constrained case.

Constrained MaxMin Angle Lemma. Among all constrained triangulations of S and L, the constrained Delaunay triangulation maximizes the minimum angle.

Extended Voronoi diagrams

Just as for ordinary Delaunay triangulations, every constrained Delaunay triangulation has a dual Voronoi diagram, but in a surface that is more complicated than the Euclidean plane. Imagine $\mathbb{R}^2$ is a sheet of paper, Σ_0, with the points of S and the line segments in L drawn on it. For each $\ell_i \in L$, we cut Σ_0 open along ℓ_i and glue another sheet Σ_i, which is also cut open along ℓ_i. The gluing is done around ℓ_i such that every traveler who crosses ℓ_i switches from Σ_0 to Σ_i and vice versa. A cross-section of the particular gluing necessary to achieve that effect is illustrated in Figure 2.4. It is not possible to do this without self-intersections in $\mathbb{R}^3$, but in $\mathbb{R}^4$ there is already sufficient space to embed the resulting surface. Call Σ_0 the *primary sheet*, and after the gluing is done we have $m = \text{card } L$ *secondary sheets* Σ_i for $1 \le i \le m$. Each secondary sheet is attached to Σ_0, but not connected to any of the other secondary sheets. For each point $x \in \mathbb{R}^2$, we now have $m + 1$ copies $x_i \in \Sigma_i$, one on each sheet.

We know what it means for two points on the primary sheet to be visible from each other. For other pairs we need a more general definition. For $i \ne 0$, points $x_0 \in \Sigma_0$ and $y_i \in \Sigma_i$ are *visible* if xy crosses ℓ_i, and ℓ_i is the first constraining line segment crossed if we traverse xy in the direction from x to y. The *distance* between points x_0 and y_i is

$$d(x_0, y_i) = \begin{cases} \|x - y\|, & \text{if } x_0, y_i \text{ are visible,} \\ \infty, & \text{otherwise.} \end{cases}$$

The new distance function is used to define the *extended Voronoi diagram*, which is illustrated in Figure 2.5. A circle that witnesses the membership of

Figure 2.4. The gap in Σ_0 represents the cut along ℓ_i. The secondary sheet Σ_i is glued to Σ_0 so that each path crossing ℓ_i switches sheets.

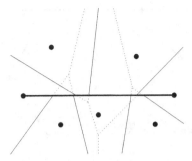

Figure 2.5. Extended Voronoi diagram dual to the constrained Delaunay triangulation in Figure 2.3. There is only one secondary sheet glued to the primary one. The solid Voronoi edges lie in the primary sheet and the dotted ones in the secondary sheet.

an edge ab in the constrained Delaunay triangulation has its center on the primary or on a secondary sheet. In either case, that center is closer to a and b than to any other point in S. This implies that the Voronoi regions of a and b meet along a nonempty common portion of their boundary. Conversely, every point on an edge of the extended Voronoi diagram is the center of a circle witnessing the membership of the corresponding edge in the constrained Delaunay triangulation.

Bibliographic notes

The idea of using a plane-sweep construction for solving two-dimensional geometric problems is almost as old as the field of computational geometry itself. It was popularized as a general algorithmic paradigm by Nievergelt and Preparata [3]. Constrained Delaunay triangulations were independently discovered by Lee and Lin [2] and by Paul Chew [1]. Extended Voronoi diagrams are from Raimund Seidel [4], who used them to construct constrained Delaunay triangulations in worst-case time $O(n \log n)$.

[1] L. P. Chew. Constrained Delaunay triangulations. In "Proc. 3rd Ann. Sympos. Comput. Geom.," 1987, 215–222.
[2] D. T. Lee and A. K. Lin. Generalized Delaunay triangulations for planar graphs. *Discrete Comput. Geom.* 1 (1986), 201–217.
[3] J. Nievergelt and F. P. Preparata. Plane-sweep algorithms for intersecting geometric figures. *Comm. ACM* **25** (1982), 739–747.
[4] R. Seidel. Constrained Delaunay triangulations and Voronoi diagrams with obstacles. In "1978–1988 Ten Years IIG," 1988, 178–191.

2.2 Delaunay refinement

This section demonstrates the use of Delaunay triangulations in constructing triangle meshes in the plane. The idea is to add new vertices until the triangulation

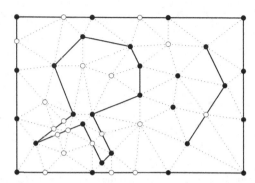

Figure 2.6. The solid vertices and edges define the input graph, and together with the hollow vertices and dotted edges they define the output triangulation.

forms a satisfying mesh. Constraining edges are covered by Delaunay edges, although forcing them into the triangulation as we did in Section 2.1 would also be possible.

The meshing problem

The general objective in mesh generation is to decompose a geometric space into elements. The elements are restricted in type and shape, and the number of elements should not be too big. We discuss a concrete version of the two-dimensional mesh generation problem.

Input. This is a polygonal region in the plane, possibly with holes and with constraining edges and vertices inside the region.
Output. This is a triangulation of the region whose edges cover all input edges and whose vertices cover all input vertices.

The graph of input vertices and edges is denoted by G, and the output triangulation is denoted by K. It is convenient to enclose G in a bounding box and to triangulate everything inside that box. A triangulation of the input region is obtained by taking a subset of the triangles. Figure 2.6 shows input and output for a particular mesh generation problem.

Triangle quality

The quality of a triangle abc is measured by its smallest angle, θ. Two alternative choices would be the largest angle and the aspect ratio. We argue that a good lower bound for the smallest angle implies good bounds for the other two expressions of quality. The largest angle is at most $\pi - 2\theta$, so if the smallest angle is bounded away from zero then the largest angle is bounded away from π. The converse is not true. The aspect ratio is the length of the longest edge, which

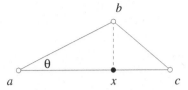

Figure 2.7. Triangle with base ac, height bx, and minimum angle θ.

we assume is ac, divided by the distance of b from ac; see Figure 2.7. Suppose the smallest angle occurs at a. Then $\|b - x\| = \|b - a\| \cdot \sin\theta$, where x is the orthogonal projection of b onto ac. The edge ab is at least as long as cb, and therefore $\|b - a\| \geq \|c - a\|/2$. It follows that

$$\frac{1}{\sin\theta} \leq \frac{\|c - a\|}{\|b - x\|} \leq \frac{2}{\sin\theta}.$$

In words, the aspect ratio is linearly related to one over the smallest angle. If θ is bounded away from zero, then the aspect ratio is bounded from above by some constant, and vice versa.

The goal is to construct K so its smallest angle is no less than some constant, and the number of triangles in K is at most some constant times the minimum. We see from the example in Figure 2.6 that a small angle between two input edges cannot possibly be resolved. A reasonable way to deal with this difficulty is to accept sharp input features as unavoidable and to isolate them so they cause no deterioration of the triangulation nearby. In this section, we assume that there are no sharp input features, and in particular that all input angles are at least $\pi/2$.

Delaunay refinement

We construct K as the Delaunay triangulation of a set of points that includes all input points. Other points are added one by one to resolve input edges that are not covered and triangles that have too small an angle.

(1) Suppose ab is a segment of an edge in G that is not covered by edges of the current Delaunay triangulation. This can only be because some of the vertices lie inside the diameter circle of ab, as in Figure 2.8. We say these vertices *encroach upon* ab, and we use function SPLIT$_1$ to add the midpoint of ab and to repair the Delaunay triangulation with a series of edge flips.

(2) Suppose a triangle abc in the current Delaunay triangulation K is skinny; that is, it has an angle less than the required lower bound. We use function SPLIT$_2$ to add the circumcenter as a new vertex, such as point x in Figure 2.9. Since its circumcircle is no longer empty, triangle abc is guaranteed to be removed by one of the edge flips used to repair the Delaunay triangulation.

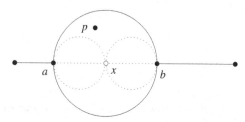

Figure 2.8. Vertex p encroaches upon segment ab. After adding the midpoint, we have two smaller diameter circles, both contained in the diameter circle of ab.

Algorithm

The first priority of the algorithm is to cover input edges, and its second priority is to resolve skinny triangles. Before starting the algorithm, we place G inside a rectangular box B. The purpose of the box is to contain the points added by the algorithm and thus prevent the perpetual growth of the meshed region. To be specific, we take B three times the size of the minimum enclosing rectangle of G. Box B has space for nine copies of the rectangle, and we place G inside the center copy. Each side of B is decomposed into three equally long edges. Refer to Figure 2.6, where for aesthetic reasons the box is drawn smaller than required but with the right combinatorics. Initially, K is the Delaunay triangulation of the input points, which includes the 12 vertices along the boundary of B.

```
loop
    while ∃ encroached segment ab do SPLIT₁(ab) endwhile;
    if  no skinny triangle left then exit endif;
    let abc ∈ K be skinny and x its circumcentre;
    x encroaches upon segments s₁, s₂, . . . , sₖ;
    if k ≥ 1 then SPLIT₁(sᵢ) for all i else  SPLIT₂(abc) endif
forever.
```

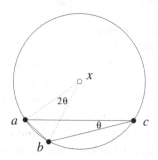

Figure 2.9. The angle $\angle axb$ is twice the angle $\angle acb$.

The choice of B implies that no circumcenter x will ever lie outside the box. This is because the initial 12 or fewer triangles next to the box boundary have nonobtuse angles opposite to boundary edges. Since the circumcircles of Delaunay triangles are empty, this implies that all circumcenters lie inside B. The algorithm maintains the nonobtuseness of angles opposing input edges and thus limits circumcenters to lie inside B.

Preliminary analysis

The behavior of the algorithm is expressed by the points it adds as vertices to the mesh. We already know that all points lie on the boundary or inside the box B, which has finite area. If we can prove that no two points are less than a positive constant 2ε apart, then this implies that the algorithm halts after adding finitely many points. To be specific, let w be the width and h the height of B. The area of the box obtained by extending B by ε on each side is $A = (w + 2\varepsilon)(h + 2\varepsilon)$. The number of points inside the box is $n \leq A/\varepsilon^2\pi$. This is because the disks with radius ε centred at the vertices of the mesh have pairwise disjoint interiors, and they are all contained in the extended box. This type of area argument is common in meshing and related to packing, as illustrated in Figure 2.10. The existence of a positive ε will be established in Section 2.3. The analysis there will refine the area argument by varying the sizes of disks with their location inside the meshing region.

In terms of running time, the most expensive activity is edge flipping used to repair the Delaunay triangulation. The expected linear bound on the number proved in Section 1.3 does not apply because points are not added in a random order. The total number of flips is less than $\binom{n}{2}$. This implies an upper bound of $O(n^2)$ on the running time, as long as the cost for adding a new vertex is at most $O(n)$.

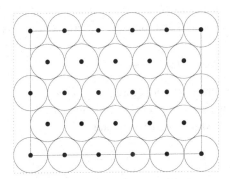

Figure 2.10. The centers of the disk are contained in the inner box, and the disks are contained in the box enlarged by the disk radius in all four directions.

Bibliographic notes

The algorithm described in this section is from Jim Ruppert [3]. Experiments suggest it achieves best results if the skinny triangles are removed in order of nondecreasing smallest angle. A predecessor of Ruppert's algorithm is the version of the Delaunay refinement method by Paul Chew [1]. That algorithm is also described in [2], where it is generalized to surfaces in three-dimensional space. The main contribution of Ruppert is a detailed analysis of the Delaunay refinement method. The gained insights are powerful enough to permit modifications of the general method that guarantee a close to optimum mesh.

[1] L. P. Chew. Guaranteed-quality triangular meshes. Report TR-98-983, Comput. Sci. Dept., Cornell Univ., Ithaca, New York, 1989.
[2] L. P. Chew. Guaranteed-quality mesh generation for curved surfaces. In "Proc. 9th Ann. Sympos. Comput. Geom.," 1993, 274–280.
[3] J. Ruppert. A Delaunay refinement algorithm for quality 2-dimensional mesh generation. *J. Algorithms* **18** (1995), 548–585.

2.3 Analysis

This section analyses the Delaunay refinement algorithm of Section 2.2. It proves an upper bound on the number of triangles generated by the algorithm and an asymptotically matching lower bound on the number of triangles that must be generated.

Local feature size

We understand the Delaunay refinement algorithm by relating its actions to the *local feature size*, defined as a map $f : \mathbb{R}^2 \to \mathbb{R}$. For a point $x \in \mathbb{R}^2$, $f(x)$ is the smallest radius r such that the closed disk with center x and radius r

(i) contains two vertices of G,
(ii) intersects one edge of G and contains one vertex of G that is not an endpoint of that edge, or
(iii) intersects two vertex disjoint edges of G.

The three cases are illustrated in Figure 2.11. Because of (i) we have $f(a) \leq \|a - b\|$ for all vertices $a \neq b$ in G. The local feature size satisfies a one-sided Lipschitz inequality, which implies continuity.

Lipschitz Condition. $|f(x) - f(y)| \leq \|x - y\|$.

Proof. To get a contradiction, assume there are points x, y with $f(x) < f(y) - \|x - y\|$. The disk with radius $f(x)$ around x is contained in the interior of the disk with radius $f(y)$ around y. We can thus shrink the disk of y while

Figure 2.11. In each case, the radius of the circle is the local feature size at x.

maintaining its nonempty intersection with two disjoint vertices or edges of G. This contradicts the definition of $f(y)$. □

Constants

The analysis of the algorithm uses two carefully chosen positive constants C_1 and C_2 such that

$$1 + \sqrt{2}C_2 \leq C_1 \leq [(C_2 - 1)/2 \sin \alpha],$$

where α is the lower bound on angles enforced by the Delaunay refinement algorithm. The constraints that correspond to the two inequalities are bounded by lines, and we have a solution if and only if the slope of the first line is greater than that of the second, $1/\sqrt{2} > 2 \sin \alpha$. Figure 2.12 illustrates the two constraints for $\alpha < \arcsin(1/2\sqrt{2}) = 20.7\ldots^\circ$. The two lines intersect at a point in the positive quadrant, and the coordinates of that point are the smallest constants C_1 and C_2 that satisfy the inequalities.

Invariants

The algorithm starts with the vertices of G and generates all other vertices in sequence. We show that, when a new vertex is added, its distance to already present vertices is not much smaller than the local feature size.

Invariants. Let p and x be two vertices such that x was added after p. If x was added by

(A) SPLIT$_1$ then $\|x - p\| \geq f(x)/C_1$,
(B) SPLIT$_2$ then $\|x - p\| \geq f(x)/C_2$.

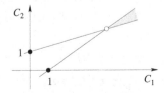

Figure 2.12. Each line bounds a half-plane of points (C_1, C_2) that satisfy one inequality. The shaded wedge contains all points that satisfy both inequalities.

Proof. We first prove (B). In this case, point x is the circumcenter of a skinny triangle abc. Let $\theta < \alpha$ at c be the smallest angle in abc, as in Figure 2.9. Assume that either a and b both belong to G or that a was added after b. We distinguish three cases, depending on how a became to be a vertex. Let L be the length of ab.

Case 1. Here a is a vertex of G. Then b is also a vertex of G and $f(a) \leq L$.

Case 2. Here a was added as the circumcenter of a circle with radius r'. Prior to the addition of a, this circle was empty, and hence $r' \leq L$. By induction, we have $f(a) \leq r' \cdot C_2$ and therefore $f(a) \leq L \cdot C_2$.

Case 3. Here a was added as the midpoint of a segment. Then $f(a) \leq L \cdot C_1$, again by induction.

Since $1 \leq C_2 \leq C_1$, we have $f(a) \leq L \cdot C_1$ in all three cases. Let $r = \|x - a\|$ be the radius of the circumcircle of abc. Using the Lipschitz Condition and $L = 2r \sin \theta$ from Figure 2.9, we get

$$f(x) \leq f(a) + r$$
$$\leq L \cdot C_1 + r$$
$$\leq 2r \cdot \sin \theta \cdot C_1 + r.$$

Since $\theta < \alpha$ and $C_2 \geq 1 + 2C_1 \cdot \sin \alpha$, we get

$$r \geq f(x)/(1 + 2C_1 \cdot \sin \alpha) \geq f(x)/C_2,$$

as required.

We use a similar argument to prove (A). In this case, x is the midpoint of a segment ab. Let $r = \|x - a\| = \|x - b\|$ be the radius of the smallest circle passing through a and b, and let p be a vertex that encroaches upon ab, as in Figure 2.8. Consider first the case in which p lies on an input edge that shares no endpoint with the input edge of ab. Then $f(x) \leq r$ by condition (iii) of the definition of local feature size. Consider second the case in which the splitting of ab is triggered by rejecting the addition of a circumcenter. Let p be this circumcenter and let r' be the radius of its circle. Since p lies inside the diameter circle of ab, we have $r' \leq \sqrt{2}r$. Using the Lipschitz Condition and induction, we get

$$f(x) \leq f(p) + r$$
$$\leq r' \cdot C_2 + r$$
$$\leq \sqrt{2}r \cdot C_2 + r.$$

Using $C_1 \geq 1 + \sqrt{2}C_2$, we get

$$r \geq f(x)/(1 + \sqrt{2}C_2) \geq f(x)/C_1,$$

as required. □

Upper bound

Invariants (A) and (B) guarantee that vertices added to the triangulation cannot get arbitrarily close to preceding vertices. We show that this implies that they cannot get close to succeeding vertices either. Recall that K is the final triangulation generated by the Delaunay refinement algorithm.

Smallest Gap Lemma. $\|a - b\| \geq f(a)/(1 + C_1)$ for all vertices $a, b \in K$.

Proof. If b precedes a then $\|a - b\| \geq f(a)/C_1 \geq f(a)/(1 + C_1)$. Otherwise, we have $\|b - a\| \geq f(b)/C_1$ and therefore

$$f(a) \leq f(b) + \|a - b\| \leq \|a - b\| \cdot (1 + C_1),$$

as claimed. □

Since vertices cannot get arbitrarily close to each other, we can use an area argument to show that the algorithm halts after adding a finite number of vertices. We relate the number of vertices to the integral of $1/f^2(x)$. Recall that B is the bounding box used in the construction of K.

Upper Bound Lemma. The number of vertices in K is at most some constant times $\int_B dx/f^2(x)$.

Proof. For each vertex a of K, let D_a be the disk with center a and radius $r_a = f(a)/(2 + 2C_1)$. By the Smallest Gap Lemma, the disks are pairwise disjoint. At least one quarter of each disk lies inside B. Therefore,

$$\int_B \frac{dx}{f^2(x)} \geq \frac{1}{4} \cdot \sum_a \int_{D_a} \frac{dx}{f^2(x)}$$

$$\geq \frac{1}{4} \cdot \sum_a \frac{r_a^2 \pi}{(f(a) + r_a)^2}$$

$$\geq \frac{1}{4} \cdot \sum_a \frac{\pi}{(3 + 2C_1)^2}.$$

This is a constant times the number of vertices. □

Two geometric results

We prepare the lower bound argument with two geometric results on triangles with angles no smaller than some constant $\alpha > 0$. Two edges of such a triangle abc cannot be too different in length, and specifically, $\|a - c\|/\|a - b\| \leq \varrho = 1/\sin\alpha$. If we have a chain of triangles connected through shared edges, the length ratio cannot exceed ϱ^t, where t is the number of triangles. Two edges sharing a common vertex are connected by the chain of triangles around that vertex. That chain cannot be longer than $2\pi/\alpha$, simply because we cannot pack more angles into 2π.

Length Ratio Lemma. The length ratio between two edges sharing a common vertex is at most $\varrho^{2\pi/\alpha}$.

The second result concerns covering a triangle with four disks, one each around the three vertices and the circumcenter. For each vertex we take a disk with radius c_0 times the length of the shortest edge. For the circumcenter we take a disk with radius $1 - c_2$ times the circumradius. For a general triangle, we can keep c_0 fixed and force c_2 as close to zero as we like, just by decreasing the angle. If angles cannot be arbitrarily small, then c_2 can also be bounded away from zero.

Triangle Cover Lemma. For each constant $c_0 > 0$ there is a constant $c_2 > 0$ such that the four disks cover the triangle.

Proof. Refer to Figure 2.13. Let R be the circumradius and ab be the shortest of the three edges. Its length is $\|a - b\| \geq 2R \cdot \sin\alpha$. The disk around a

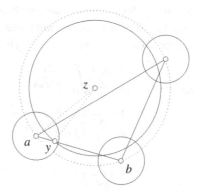

Figure 2.13. The disks constructed for a triangle and its three vertices cover the triangle.

covers all points at distance at most $c_0 \cdot \|a - b\|$ from a, and we assume without loss of generality that $c_0 < 1/2$. The distance between the circumcenter, z, and the point $y \in ab$ at distance $c_0 \cdot \|a - b\|$ from a is

$$\|y - z\| < \sqrt{R^2 - c_0^2 \|a - b\|^2}$$
$$\leq \sqrt{R^2 \cdot \left(1 - 4c_0^2 \cdot \sin^2 \alpha\right)}$$
$$< R \cdot \left(1 - 2c_0^2 \cdot \sin^2 \alpha\right).$$

All other points on triangle edges not covered by disks around a, b, c are at most that distance from z. Since c_0 and α are positive constants, $c_2 = 2c_0^2 \cdot \sin^2 \alpha$ is also a positive constant. □

Lower bound

The reason for picking the disk of radius $(1 - c_2)R$ around the circumcenter is that for a point x inside this disk, the local feature size cannot be arbitrarily small. In particular, it cannot be smaller than the distance from the circumcircle times the cosine of half the smallest angle, $f(x) \geq c_2 R \cdot \cos(\alpha/2)$. To get a similar result for disks around vertices, let L be the length of the shortest edge incident to a vertex a. The local feature size of a is at least $L \cdot \sin \alpha$. By choosing $c_0 = \sin \alpha/2$ we get $f(a) \geq 2c_0 L$, and therefore $f(x) \geq f(a) - \|a - x\| \geq c_0 L$ for every point x inside the disk with radius $c_0 L$ around a.

We use these observations to show that any algorithm that constructs triangles with angles no smaller than some constant $\alpha > 0$ generates at least some constant times the integral of $1/f^2(x)$ many vertices. It follows that the algorithm in Section 2.2 constructs meshes with asymptotically minimum size.

Lower Bound Lemma. If K is a triangle mesh of G with all angles larger than α, then the number of vertices is at least some constant times $\int_B dx/f^2(x)$.

Proof. Around each vertex $a \in K$ draw a disk with radius equal to $\sin \alpha/2$ times the length of the shortest incident edge. Let $c_0 = \sin \alpha \cdot \varrho^{\pi/\alpha}/2$ and use the Triangle Cover Lemma to pick a matching constant $c_2 > 0$. For each triangle $abc \in K$, draw the disk with radius $1 - c_2$ times the circumradius around the circumcenter. Each triangle is covered by its four disks, which implies that the mesh is covered by the collection of disks.

For each disk D_i in the collection, let f_i be the minimum local feature size at any point $x \in D_i$. By what we said earlier, that minimum is at least some constant fraction of the radius of D_i, $f_i \geq r_i/C$. Given that the disks cover the mesh, we have

$$\int_B \frac{dx}{f^2(x)} \leq \sum_i \int_{D_i} \frac{dx}{f^2(x)}$$

$$\leq \sum_i \frac{r_i^2 \pi}{f_i^2}$$

$$\leq \sum_i C^2 \pi.$$

The number of triangles is less than twice the number of vertices, which we denote as n. Hence,

$$n \geq \sum_i \frac{1}{3} \geq \frac{1}{3C^2\pi} \int_B \frac{dx}{f^2(x)},$$

as claimed.   □

Bibliographic notes

The idea of using the local feature size function in the analysis of the Delaunay refinement algorithm is from Jim Ruppert. The details of the analysis left out in the journal publication [3] can be found in the technical report [2]. Bern, Eppstein, and Gilbert [1] show that the same technical result (constant minimum angle and constant times minimum number of triangles) can also be achieved by using quad trees. Experimentally, the approach with Delaunay triangulations seems to generate meshes with fewer and nicer triangles. One reason for the better performance might be the absence of any directional bias from Delaunay triangulations.

[1] M. Bern, D. Eppstein, and J. Gilbert. Provably good mesh generation. *J. Comput. Syst. Sci.* **48** (1994), 384–409.
[2] J. Ruppert. A new and simple algorithm for quality 2-dimensional mesh generation. Report UCB/CSD 92/694, Comput. Sci. Div., Univ. California, Berkeley, California, 1992.
[3] J. Ruppert. A Delaunay refinement algorithm for quality 2-dimensional mesh generation. *J. Algorithms* **18** (1995), 548–585.

Exercise collection

The credit assignment reflects a subjective assessment of difficulty. A typical question can be answered by using knowledge of the material combined with some thought and analysis.

1. **Acute triangles** (one credit). An *acute* triangle has all three angles less than $\pi/2$. Note that a Delaunay triangle abc is acute if and only if the dual Voronoi

vertex is contained in the interior of *abc*. Show that a triangulation K all of whose triangles are acute is necessarily the Delaunay triangulation of its vertex set.

2. **Gabriel graph** (one credit). Let S be a finite set in the plane. Let $a, b \in S$ and consider the smallest circle C_{ab} that passes through a and b. The edge *ab* belongs to the *Gabriel graph* G of S if C_{ab} is empty and a, b are the only two points of S that lie on the circle.

   (i) Show that $ab \in G$ if and only if *ab* belongs to the Delaunay triangulation and the opposite angles in the one or two triangles that share *ab* are less than $\pi/2$.

   (ii) Show that $ab \in G$ if and only if *ab* crosses the dual Voronoi edge $V_a \cap V_b$.

3. **Voronoi diagram for line segments** (three credits). Consider a set L of n pairwise disjoint line segments in the plane. The *distance* of a point $x \in \mathbb{R}^2$ from $ab \in L$ is the minimum Euclidean distance from any point on *ab*. The *Voronoi region* of *ab* is then the set of points x for which *ab* is no further than any other line segment in L.

   (i) Prove that the Voronoi region of every line segment in L is connected.

   (ii) Show that the edges of the Voronoi regions are pieces of lines and parabolas.

   (iii) Prove that the number of Voronoi edges is at most some constant times n.

4. **Noncrossing edges** (two credits). Let S be a finite set of points in $\mathbb{R}^2$ such that no three lie on a common line and no four lie on a common circle. We say two edges *ab* and *cd cross* if they share a common interior point, int $ab \cap$ int $cd \neq \emptyset$. Let L be a set of pairwise noncrossing line segments with endpoints in S.

   (i) Prove that no two edges of the Delaunay triangulation of S cross.

   (ii) Prove that no two edges of the constrained Delaunay triangulation of S and L cross.

5. **Surrounded Voronoi vertices** (three credits). Consider the Delaunay triangulation of a finite set of points in $\mathbb{R}^2$. Let D be a subset of the Delaunay triangles, with boundary B consisting of all edges that belong to exactly one triangle in D. We call D *protected* if $abc \in D$ and $ab \in B$ implies the angle at c is nonobtuse. Prove that all Voronoi vertices dual to triangles in a protected subset of triangles, D, lie in the regions covered by D.

3
Combinatorial topology

The primary purpose of this chapter is the introduction of standard topological language to streamline our discussions on triangulations and meshes. We will spend most of the effort to develop a better understanding of space, how it is connected, and how we can decompose it. The secondary purpose is the construction of a bridge between continuous and discrete concepts in geometry. The idea of a continuous and possibly even differential world is close to our intuitive understanding of physical phenomena, while the discrete setting is natural for computation. Section 3.1 introduces simplicial complexes as a fundamental discrete representation of continuous space. Section 3.2 talks about refining complexes by decomposing simplices into smaller pieces. Section 3.3 describes the topological notion of space and the important special case of manifolds. Section 3.4 discusses the Euler characteristic of a triangulated space.

3.1 Simplicial complexes

We use simplicial complexes as the fundamental tool to model geometric shapes and spaces. They generalize and formalize the somewhat loose geometric notions of a triangulation. Because of their combinatorial nature, simplicial complexes are perfect data structures for geometric modeling algorithms.

Simplices

A finite collection of points is *affinely independent* if no affine space of dimension i contains more than $i + 1$ of the points, and this is true for every i. A *k-simplex* is the convex hull of a collection of $k + 1$ affinely independent points, $\sigma = \operatorname{conv} S$. The *dimension* of σ is $\dim \sigma = k$. In $\mathbb{R}^d$, the largest number of affinely independent points is $d + 1$, and we have simplices of dimension $-1, 0, \ldots, d$. The (-1)-simplex is the empty set. Figure 3.1 shows the four

0   1   2   3

Figure 3.1. A 0-simplex is a point or vertex, a 1-simplex is an edge, a 2-simplex is a triangle, and a 3-simplex is a tetrahedron.

types of nonempty simplices in $\mathbb{R}^3$. The convex hull of any subset $T \subseteq S$ is again a simplex. It is a subset of conv S and called a *face* of σ, which is denoted as $\tau \leq \sigma$. If dim $\tau = \ell$ then τ is called an *ℓ-face*. $\tau = \emptyset$ and $\tau = \sigma$ are *improper* faces, and all others are *proper* faces of σ. The number of ℓ faces of σ is equal to the number of ways we can choose $\ell + 1$ from $k + 1$ points, which is $\binom{k+1}{\ell+1}$. The total number of faces is

$$\sum_{\ell=-1}^{k} \binom{k+1}{\ell+1} = 2^{k+1}.$$

Simplicial complexes

A *simplicial complex* is the collection of faces of a finite number of simplices, any two of which are either disjoint or meet in a common face. More formally, it is a collection K such that

(i) $\sigma \in K \land \tau \leq \sigma \Longrightarrow \tau \in K$, and
(ii) $\sigma, \upsilon \in K \Longrightarrow \sigma \cap \upsilon \leq \sigma, \upsilon$.

Note that $\emptyset$ is a face of every simplex and thus belongs to K by Condition (i). Condition (ii) therefore allows for the possibility that σ and υ be disjoint. Figure 3.2 shows three sets of simplices that each violate one of the two conditions and therefore do not form complexes. A *subcomplex* is a subset that is a simplicial complex itself. Observe that every subset of a simplicial complex

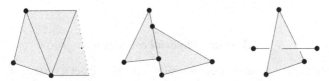

Figure 3.2. To the left, we are missing an edge and two vertices. In the middle, the triangles meet along a segment that is not an edge of either triangle. To the right, the edge crosses the triangle at an interior point.

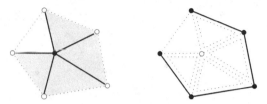

Figure 3.3. Star and link of a vertex. To the left, the solid edges and shaded triangles belong to the star of the solid vertex. To the right, the solid edges and vertices belong to the link of the hollow vertex.

satisfies Condition (ii). To enforce Condition (i), we may add faces and simplices to the subset. Formally, the *closure* of a subset $L \subseteq K$ is the smallest subcomplex that contains L,

$$\text{Cl}\, L = \{\tau \in K \mid \tau \leq \sigma \in L\}.$$

A particular subcomplex is the *i-skeleton* that consists of all simplices $\sigma \in K$ whose dimension is i or less. The *vertex set* is Vert $K = \{\sigma \in K \mid \dim \sigma = 0\}$, which is the 0-skeleton minus the (-1)-simplex. The *dimension* of K is the largest dimension of any simplex, $\dim K = \max\{\dim \sigma \mid \sigma \in K\}$. If $k = \dim K$ then K is a *k-complex*.

Stars and links

We use special subsets to talk about the local structure of a simplicial complex. These subsets may or may not be closed. The *star* of a simplex τ consists of all simplices that contain τ, and the *link* consists of all faces of simplices in the star that do not intersect τ,

$$\text{St}\, \tau = \{\sigma \in K \mid \tau \leq \sigma\},$$

$$\text{Lk}\, \tau = \{\sigma \in \text{Cl}\, \text{St}\, \tau \mid \sigma \cap \tau = \emptyset\}.$$

Figure 3.3 illustrates this definition by showing the star and the link of a vertex in a 2-complex. The star is generally not closed, but the link is always a simplicial complex.

Abstract simplicial complexes

By substituting the *set* of vertices for each simplex, we get a system of subsets of the vertex set. In doing so, we throw away the geometry of the simplices and focus on the combinatorial structure. Formally, a finite system A of finite sets is an *abstract simplicial complex* if $\alpha \in A$ and $\beta \subseteq \alpha$ implies $\beta \in A$. This

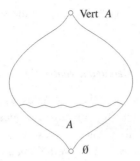

Figure 3.4. The onion is the power set of Vert A. The area below the waterline is an abstract simplicial complex.

requirement is similar to Condition (i) for geometric simplicial complexes. A set $\alpha \in A$ is an *(abstract) simplex* and its dimension is $\dim \alpha = \operatorname{card} \alpha - 1$. The *vertex set* of A is $\operatorname{Vert} A = \bigcup A = \bigcup_{\alpha \in A} \alpha$.

Observe that A is a subsystem of the power set of Vert A. We can therefore think of it as a subcomplex of an n-simplex, where $n + 1 = \operatorname{card} \operatorname{Vert} A$. This view is expressed in the picture of an abstract simplicial complex shown as Figure 3.4. The concepts of face, subcomplex, closure, star, and link extend straightforwardly from geometric to abstract simplicial complexes.

Posets

The set system together with the inclusion relation forms a partially ordered set, or poset, denoted as $(A, \subseteq)$. Posets are commonly drawn by using Hasse diagrams, where sets are nodes, smaller sets are below larger sets, and inclusions are edges. Figure 3.5 shows the Hasse diagrams of simplices with dimensions 0 to 3. Implied inclusions are usually not drawn.

Here is a recursive way to construct the Hasse diagram of a k-simplex α. First draw the Hasse diagrams for two $(k - 1)$-simplices. One is the diagram of a $(k - 1)$-face β of α, and the other is the diagram for the star of the vertex

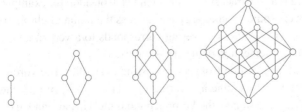

Figure 3.5. From left to right, the poset of a vertex, an edge, a triangle, and a tetrahedron.

$u \in \alpha - \beta$. Finally, connect every simplex γ in the star of u with the simplex $\gamma - \{u\}$ in the closure of β.

Geometric realization

We can think of an abstract simplicial complex as an abstract version of a geometric simplicial complex. To formalize this idea, we define a *geometric realization* of an abstract simplicial complex A as a simplicial complex K together with a bijection $\varphi : \text{Vert } A \to \text{Vert } K$ such that $\alpha \in A$ if and only if $\text{conv } \varphi(\alpha) \in K$. A is sometimes called an *abstraction* of K.

Given A, we can ask for the smallest number of dimensions that allow a geometric realization. For example, graphs are one-dimensional abstract simplicial complexes and can always be realized in $\mathbb{R}^3$. Two dimensions are sometimes but not always sufficient. This result generalized to k-dimensional abstract simplicial complexes. They can always be realized in $\mathbb{R}^{2k+1}$ and sometimes $\mathbb{R}^{2k}$ does not suffice. To prove the sufficiency of the claim, we show that the k-skeleton of every n-simplex can be realized in $\mathbb{R}^{2k+1}$. Map the $n + 1$ vertices to points in general position in $\mathbb{R}^{2k+1}$. Specifically, we require that any $2k + 2$ of the points are affinely independent. Two simplices σ and υ of the k-skeleton have a total of at most $2(k + 1)$ vertices, which are therefore affinely independent. In other words, σ and υ are faces of a common simplex of dimension at most $2k + 1$. Hence, $\sigma \cap \upsilon$ is a common face of both.

Nerves

A convenient way to construct abstract simplicial complexes starts from an arbitrary finite set. The *nerve* of such a set C is the system of subsets with nonempty intersection,

$$\text{Nrv } C = \left\{ \alpha \subseteq C \mid \bigcap \alpha \neq \emptyset \right\}.$$

If $\beta \subseteq \alpha$ then $\bigcap \alpha \subseteq \bigcap \beta$. Hence $\alpha \in \text{Nrv } C$ implies $\beta \in \text{Nrv } C$, which shows that the nerve is an abstract simplicial complex. Consider, for example, the case where C covers some geometric space, such as the union of elliptic regions in Figure 3.6. Every set in the covering corresponds to a vertex, and $k + 1$ sets with nonempty intersection define a k-simplex.

We have seen an example of such a construction earlier. The Voronoi regions of a finite set $S \subseteq \mathbb{R}^2$ define a covering $C = \{V_a \mid a \in S\}$ of the plane. When general position is assumed, the Voronoi regions meet in pairs and in triplets, but not in quadruplets. The nerve contains abstract vertices, edges, and triangles, but

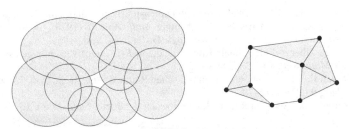

Figure 3.6. A covering with eight sets to the left and a geometric realization of its nerve to the right. The sets meet in triplets but not in quadruplets, which implies that the nerve is two dimensional.

no abstract tetrahedra. Consider the function $\varphi : C \to \mathbb{R}^2$ that maps a Voronoi region to its generator, $\varphi(V_a) = a$. This function defines a geometric realization of Nrv C, namely

$$D = \{\operatorname{conv} \varphi(\alpha) | \alpha \in \operatorname{Nrv} C\}.$$

This is the Delaunay triangulation of S. What happens if the points in S are not in general position? If $k + 1 \geq 4$ Voronoi regions have a nonempty common intersection, then Nrv C contains the corresponding abstract k-simplex. So instead of making a choice among the possible triangulations of the $(k + 1)$-gon, the nerve takes all possible triangulations together and interprets them as subcomplexes of a k-simplex. The disadvantage of this method is of course that a k-simplex for $k \geq 3$ cannot be realized in $\mathbb{R}^2$.

Bibliographic notes

The concept of a nerve was introduced to combinatorial topology in 1928 by Alexandrov [1]. During the first half of the twentieth century, combinatorial topology was a flourishing field within mathematics. We refer to the work of Paul Alexandrov [2] as a comprehensive text originally published as a series of three books. This text roughly coincides with a fundamental reorganization of the field triggered by a variety of technical results in topology. One of the successors of combinatorial topology is modern algebraic topology, where the emphasis shifts from combinatorial to algebraic structures.

We have seen that every k-complex can be geometrically realized in $\mathbb{R}^{2k+1}$. Examples of k-complexes that require $2k + 1$ dimensions are provided by Flores [3] and independently by van Kampen [4]. One such example is the k-skeleton of the $(2k + 2)$-simplex. For $k = 1$, this is the complete graph of five vertices, which is one of the two obstructions of graph planarity identified by Kuratowski [5].

[1] P. S. Alexandrov. Über den allgemeinen Dimensionsbegriff und seine Beziehungen zur elementaren geometrischen Anschauung. *Math. Ann.* **98** (1928), 617–635.

[2] P. S. Alexandrov. *Combinatorial Topology.* Dover, New York, 1956.

[3] A. Flores. Über n-dimensionale Komplexe die in R_{2n+1} selbstverschlungen sind. *Ergeb. Math. Koll.* **6** (1933/34), 4–7.

[4] E. R. van Kampen. Komplexe in euklidischen Räumen. *Abh. Math. Sem. Univ. Hamburg* **9** (1933), 72–78.

[5] K. Kuratowski. Sur le problème des courbes en topologie. *Fund. Math.* **15** (1930), 271–283.

3.2 Subdivision

Subdividing or refining a simplicial complex means decomposing its simplices into pieces. This section discusses two ways to subdivide systematically. Both ways are based on describing points by using barycentric coordinates, which are introduced first.

Barycentric coordinates

Let S be a finite set of points p_i in $\mathbb{R}^d$. An *affine combination* is a point $x = \sum_i \gamma_i p_i$ with $\sum_i \gamma_i = 1$. The *affine hull* is the set of affine combinations and is denoted as aff S. It is the smallest affine subspace that contains S. A *convex combination* is an affine combination with $\gamma_i \geq 0$ for all i. The *convex hull* is the set of convex combinations and is denoted as conv S. In general, the γ_i are not unique, but if the p_i are affinely independent then they are. Indeed, if the $k + 1 = \text{card } S$ points are affinely independent, then the affine hull has dimension k. There are $k + 1$ coefficients, and the requirement they sum to 1 reduces the degree of freedom to k, just enough for k dimensions.

Assume the points in S are affinely independent; hence $\sigma = \text{conv } S$ is a k-simplex. The *barycentric coordinates* of a point $x \in \text{conv } S$ are the coefficients γ_i such that $x = \sum_i \gamma_i p_i$. They are all nonnegative and they add to 1. For a particular realization of $\sigma = \text{conv } S$ in $\mathbb{R}^{k+1}$, the barycentric coordinates are exactly the Cartesian coordinates. This realization is the *standard k-simplex*, which is the convex hull of the endpoints of the $k + 1$ unit vectors, as illustrated in Figure 3.7. It is also the intersection of the hyperplane $\sum_i \gamma_i x_i = 1$ with

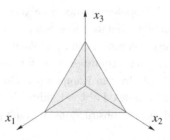

Figure 3.7. The standard triangle connects points $(1, 0, 0)$, $(0, 1, 0)$, and $(0, 0, 1)$.

the orthant of points with nonnegative coordinates. The *boundary* consists of points that have at least one barycentric coordinate equal to 0. The *interior* is the rest of the simplex, $\operatorname{int}\sigma = \sigma - \operatorname{bd}\sigma$. The *barycenter* is the point with all $k + 1$ barycentric coordinates the same, namely equal to $1/(k + 1)$. It is also known as the centroid and the center of mass.

Barycentric subdivision

A *subdivision* of a complex K is a complex L such that every simplex $\tau \in L$ is contained in a simplex $\sigma \in K$. In other words, every $\sigma \in K$ is the union of simplices in L. We describe a particular subdivision obtained by connecting the barycenters of simplices. Consider first a k-simplex, σ, and take the collection of barycenters of all faces. We connect every subset of barycenters that come from a sequence of proper faces. Figure 3.8 illustrates the construction for $k = 2$. A k-simplex is decomposed into smaller k-simplices, each the set of points with barycentric coordinates $\gamma_{i_0} \leq \gamma_{i_1} \leq \cdots \leq \gamma_{i_k}$ for some fixed permutation of the $k + 1$ indices. It follows there are $(k + 1)!$ k-simplices in the decomposition. The implied subdivision of each face is the barycentric subdivision of that face. We can therefore define the *barycentric subdivision* of K as the one obtained by subdividing every simplex as described.

There is a refreshingly different abstract description of the same construction. Let A be the abstraction of K viewed as a partially ordered set. A *chain* is a properly nested sequence of nonempty abstract simplices, $\alpha_0 \subset \alpha_1 \subset \cdots \subset \alpha_i$. The *order complex* is obtained by taking the nonempty abstract simplices of A as vertices and the chains of A as abstract simplices. Every subchain of a chain is again a chain, which implies that the order complex is an abstract simplicial complex. It is the abstraction of the barycentric subdivision of K.

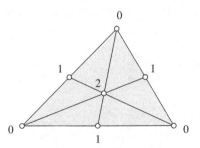

Figure 3.8. Barycentric subdivision of a triangle. Each barycenter is labeled with the dimension of the corresponding face of the triangle.

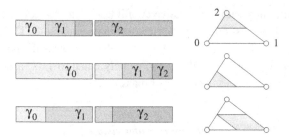

Figure 3.9. Three generic 3-divisions.

Dividing an interval

The barycentric subdivision has nice structural properties but terrible numerical behavior. We prepare the introduction of a subdivision that preserves angles to the extent this is possible. Let $x = \sum_{i=0}^{k} \gamma_i p_i$. Since $\sum_i \gamma_i = 1$, we can associate x with the division of $[0, 1]$ into $k + 1$ pieces of lengths $\gamma_0, \gamma_1, \ldots, \gamma_k$. We call this a $(k + 1)$-*division*. The map between points of σ and $(k + 1)$-divisions of $[0, 1]$ is a bijection. We use this observation and subdivide σ by distinguishing different ways to divide $[0, 1]$.

As an example, consider the case in which σ is a triangle. Suppose we cut $[0, 1]$ into two halves, and we distinguish divisions depending on which halves contain their dividing points. The three generic possibilities are shown in Figure 3.9. The first case is defined by $\gamma_2 \geq \frac{1}{2}$. There is a bijection to the 3-divisions of $[0, \frac{1}{2}]$, and therefore to the points of a triangle. Similarly, the second case is defined by $\gamma_0 \geq \frac{1}{2}$, and again we get a bijection to the points of a triangle. The third case defined by $\gamma_0, \gamma_2 \leq \frac{1}{2}$ is more interesting. We have two independent 2-divisions corresponding to a 1-simplex or edge each. The set of all pairs of 2-divisions corresponds to the Cartesian product of the two edges, which is a rhombus. To subdivide the rhombus, we stack the two intervals and distinguish the case in which the upper divider precedes the lower divider from the one in which is succeeds it. Both cases are illustrated in Figure 3.10. We make the stack into a single interval by extending the dividers over both rows.

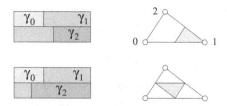

Figure 3.10. Two pairs of generic 2-divisions.

In each case, we get a bijection to the set of 3-divisions and hence to the points of a triangle. The subdivision we just described starts by cutting [0, 1] into two halves. In general, we cut it into $j \geq 1$ equal intervals, which are all stacked.

Edgewise subdivision

The method of subdividing a triangle can be generalized to d-simplices in a fairly straightforward manner. Keep in mind that the $(k + 1)$-divisions of [0, 1] form a k-simplex, and the independent division of intervals corresponds to taking the Cartesian product.

Take a $(k + 1)$-division of [0, 1]. Cut [0, 1] into j equally long intervals and stack them up, one on top of the next. At this moment, we have each interval divided into pieces, and each piece has a *color*, which is the index of the corresponding vertex. Extend the dividers through the entire stack. Assuming all dividers are different, we now have each interval divided into $k + 1$ pieces, and each piece still has its original color. This state of the construction is illustrated in Figure 3.11. To remove the length information, we transform the stack to a matrix of colors, and we refer to it as a *color scheme*. For example, the color scheme of the stack in Figure 3.11 is

$$\begin{bmatrix} 0 & 0 & 0 & 1 \\ 1 & 2 & 2 & 2 \\ 2 & 2 & 3 & 3 \end{bmatrix}.$$

If we read a color scheme read row by row like text, we get a nondecreasing sequence of colors. The other defining property is that no two columns are the same. In the generic case, there are $k + 1$ columns, and two contiguous columns differ by one color. In general, the matrix corresponds to a simplex whose dimension is the number of columns minus one. Each column corresponds to a vertex of that simplex. Specifically, the vertex that corresponds to the column with colors $i_1 \leq i_2 \leq \cdots \leq i_j$ is the average of the corresponding vertices, namely $\sum_{\ell=1}^{j} p_{i_\ell}/j$. For a fixed $j \geq 1$, the *edgewise subdivision* consists of all simplices that correspond to color schemes with j rows and $k + 1$ colors.

0	0	0	1
1	2	2	2
2	2	3	3

Figure 3.11. Stack of 4-division, cut into three equal intervals.

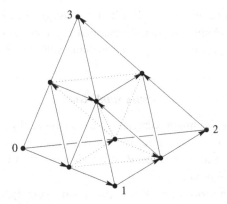

Figure 3.12. 8-division of a tetrahedron with shape vectors indicated by arrowheads.

Example

Consider the edgewise subdivision of a tetrahedron for $j = 2$. There are eight generic color schemes, namely

$$\begin{bmatrix} 0 & 0 & 0 & 0 \\ 0 & 1 & 2 & 3 \end{bmatrix}, \begin{bmatrix} 0 & 0 & 0 & 1 \\ 1 & 2 & 3 & 3 \end{bmatrix}, \begin{bmatrix} 0 & 0 & 1 & 1 \\ 1 & 2 & 2 & 3 \end{bmatrix},$$

$$\begin{bmatrix} 0 & 0 & 1 & 2 \\ 2 & 3 & 3 & 3 \end{bmatrix}, \begin{bmatrix} 0 & 1 & 1 & 1 \\ 1 & 1 & 2 & 3 \end{bmatrix}, \begin{bmatrix} 0 & 1 & 1 & 2 \\ 2 & 2 & 3 & 3 \end{bmatrix},$$

$$\begin{bmatrix} 0 & 1 & 2 & 2 \\ 2 & 2 & 2 & 3 \end{bmatrix}, \begin{bmatrix} 0 & 1 & 2 & 3 \\ 3 & 3 & 3 & 3 \end{bmatrix}.$$

They divide the tetrahedron into four tetrahedra near the vertices and four tetrahedra dividing the remaining octahedron, as shown in Figure 3.12. Note that the way the tetrahedron is subdivided depends on the ordering of the four original vertices. The distinguishing feature is the diagonal of the octahedron used in the subdivision. It corresponds to the two-by-two color scheme with colors 0, 1, 2, 3. The diagonal is therefore the edge connecting the midpoints of $p_0 p_2$ and $p_1 p_3$.

Shape vectors

In Figure 3.12, the four tetrahedra next to the original vertices are just scaled-down copies of the original tetrahedron. They are congruent, but the four tetrahedra subdividing the octahedron have usually different shape. The key to understanding the types of simplices that arise are the vectors connecting contiguous vertices of the original tetrahedron. Let $p_0, p_1, \ldots, p_k$ be an ordering of the vertices of a k-simplex σ. We have k *shape vectors*, namely $p_i - p_{i-1}$ for $1 \le i \le k$, which form a path from p_0 to p_k.

Consider the shape vectors of one of the k-simplices in the subdivision. We get the ordering of its vertices by reading the columns of the corresponding color scheme from left to right. Two contiguous columns differ in one row, and in that row the color increases by one. The corresponding shape vector is therefore just a scaled-down copy of one of the original shape vectors. After rescaling, the shape vectors of the k-simplex in the subdivision and the original k-simplex are the same, but they may come in a different order. There are k-shape vectors and thus $k!$ possible orderings. Reversing the ordering is the same as centrally reflecting the simplex. Recall that two simplices are congruent if one is obtained from the other by translation and possibly central reflection. It follows that the number of congruence classes among the k-simplices in the subdivision is at most $k!/2$, no matter how large j is. In the generic case, the number of congruence classes is exactly $k!/2$.

Bibliographic notes

The barycentric subdivision is a popular tool in applications in which numerical behavior is not important. The corresponding abstract order complex is a notion that goes back to Alexandrov [1]. The edgewise subdivision combines elegant combinatorial structure with good numerical behavior. It was independently discovered several times in the literature, and possibly the oldest reference is the paper by Freudenthal [3]. The three-dimensional case is particularly relevant in mesh generation [4]. The exposition in this section is based on the recent paper by Edelsbrunner and Grayson [2], which introduces the elementary framework of interval division and proves a variety of symmetry properties of the subdivision.

[1] P. S. Alexandrov. Diskrete Räume. *Mat. Sb.* **2** (1937), 501–518.
[2] H. Edelsbrunner and D. R. Grayson. Edgewise subdivision of a simplex. *Discrete Comput. Geom.* **24** (2000), 707–719.
[3] H. Freudenthal. Simplizialzerlegung von beschränkter Flachheit. *Ann. Math.* **43** (1942), 580–582.
[4] A. Liu and B. Joe. Quality local refinement of tetrahedral meshes based on 8-subtetrahedron subdivision. *Math. Comput.* **65** (1996), 1183–1200.

3.3 Topological spaces

The topological notions in this book are predominantly combinatorial. To understand the connection to continuous phenomena, we need but a few basic concepts from point set topology. This section introduces topological spaces, homeomorphisms, triangulations, and manifolds.

Topology

The most fundamental concept in point set topology is a *topological space*, which is a point set $\mathbb{X}$ together with a system X of subsets $A \subseteq \mathbb{X}$ that satisfies

  (i) $\emptyset, \mathbb{X} \in X$,
  (ii) $Z \subseteq X$ implies $\bigcup Z \in X$, and
 (iii) $Z \subseteq X$ and Z finite implies $\bigcap Z \in X$.

The system X is a *topology* and its sets are the *open sets* in $\mathbb{X}$. This definition is exceedingly general and rather nonintuitive, but with time we will get a better feeling for what a topological space really is. The most important example for us is the *d-dimensional Euclidean space*, denoted as $\mathbb{R}^d$. We use the Euclidean distance function to define an *open ball* as the set of all points closer than some given distance from a given point. The topology of $\mathbb{R}^d$ is the system of open sets, where each open set is a union of open balls.

All other topological spaces in this book are subsets of $\mathbb{R}^d$. A *topological subspace* of the pair $\mathbb{X}$, X is a subset $\mathbb{Y} \subseteq \mathbb{X}$ together with the *subspace topology* consisting of all intersections between $\mathbb{Y}$ and open sets, $Y = \{\mathbb{Y} \cap A \mid A \in X\}$. An example is the *d-ball*, defined as the set of points at distance 1 or less from the origin,

$$\mathbb{B}^d = \{x \in \mathbb{R}^d \mid \|x\| \le 1\}.$$

Its open sets are the intersections of $\mathbb{B}^d$ with open sets in $\mathbb{R}^d$. Note that an open set in $\mathbb{B}^d$ is not necessarily open in $\mathbb{R}^d$.

Homeomorphisms

Topological spaces are considered the same or of the same type if they are connected the same way. What it means to be connected the same way still has to be defined. There are several possibilities, and the most important one is based on homeomorphisms, which are functions between topological spaces. Such a function is *continuous* if the preimage of every open set is open, and if it is continuous it is referred to as a *map*. A *homeomorphism* is a function $f : \mathbb{X} \to \mathbb{Y}$ that is bijective, continuous, and has a continuous inverse. If a homeomorphism exists then $\mathbb{X}$ and $\mathbb{Y}$ are *homeomorphic*, and this is denoted as $\mathbb{X} \approx \mathbb{Y}$. If we want to stress that $\approx$ is an equivalence relation, we say that

Figure 3.13. From left to right, the open interval, the closed interval, the half-open interval, the circle, a bifurcation.

$\mathbb{X}$ and $\mathbb{Y}$ are *topologically equivalent* or that they have the same *topological type*. Figure 3.13 shows five examples of one-dimensional spaces with pairwise different topological types. For another example, consider the open unit disk, which is the set of points in $\mathbb{R}^2$ at distance less than one from the origin. This disk can be stretched over the entire plane. Define $f(x) = x/(1 - \|x\|)$, which maps x to the point on the same radiating half-line at the original distance times $1/(1 - \|x\|)$ from the origin. Function f is bijective and continuous, and its inverse is continuous. It follows that the open disk is homeomorphic to $\mathbb{R}^2$. More generally, every open k-dimensional ball is homeomorphic to $\mathbb{R}^k$.

Triangulation

The meaning of the term changes from one area to another. In geometry, there is no generally agreed upon definition, but it usually means a simplicial complex. In topology, a triangulation has a precise meaning, and that meaning is similar to the idea of a mesh that gives combinatorial structure to space.

Let K be a simplicial complex in $\mathbb{R}^d$. Its *underlying space* is the union of its simplices together with the subspace topology inherited from $\mathbb{R}^d$,

$$|K| = \{x \in \mathbb{R}^d \mid x \in \sigma \in K\}.$$

A *polyhedron* is the underlying space of a simplicial complex. We can think of K as a combinatorial structure imposed on $|K|$. There are others. Using homeomorphisms, we can impose the same structure on spaces that are not polyhedra. A *triangulation* of a topological space $\mathbb{X}$ is a simplicial complex K whose underlying space is homeomorphic to $\mathbb{X}$, $|K| \approx \mathbb{X}$. The space $\mathbb{X}$ is *triangulable* if it has a triangulation. An example is the triangulation of the closed disk $\mathbb{B}^2$ with nine triangles shown in Figure 3.14.

Manifolds

Manifolds are particularly nice topological spaces. They are defined locally. A *neighborhood* of a point $x \in \mathbb{X}$ is an open set that contains x. There are many neighborhoods, and usually it suffices to take one that is sufficiently

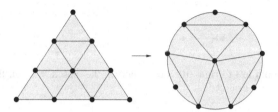

Figure 3.14. Triangulation of the closed disk. The homeomorphism maps each vertex, edge, and triangle to a homeomorphic subset of the disk.

small. A topological space $\mathbb{X}$ is a *k-manifold* if every $x \in \mathbb{X}$ has a neighborhood homeomorphic to $\mathbb{R}^k$. It is probably more intuitive to mentally substitute a small open k-ball for $\mathbb{R}^k$, but this makes no difference because the two are homeomorphic.

A simple example of a k-manifold is the *k-sphere*, which is the set of points at unit distance from the origin in the $(k+1)$-dimensional Euclidean space,

$$\mathbb{S}^k = \{x \in \mathbb{R}^{k+1} \mid \|x\| = 1\}.$$

Examples are shown in Figure 3.15. The smallest triangulation of $\mathbb{S}^k$ is the boundary complex of a $(k+1)$-simplex σ. To construct a homeomorphism, we place σ so it contains the origin in its interior, and we centrally project every point of σ's boundary onto the sphere.

Manifolds with boundary

All points of a manifold have the same neighborhood. We get a more general class of spaces if we allow two types of neighborhoods. The second type is half an open ball. Again we can stretch that space, this time over half the Euclidean space of the same dimension. Formally, the *k-dimensional half-space* is

$$\mathbb{H}^k = \{x = (x_1, x_2, \ldots, x_k) \in \mathbb{R}^k \mid x_1 \geq 0\}.$$

A space $\mathbb{X}$ is a *k-manifold with boundary* if every point $x \in \mathbb{X}$ has a neighborhood homeomorphic to $\mathbb{R}^k$ or to $\mathbb{H}^k$. The *boundary* is the set of points with a

Figure 3.15. The 0-sphere is a pair of points, the 1-sphere is a circle, and the 2-sphere is what we usually call a sphere.

Figure 3.16. The 0-ball is a point, the 1-ball is a closed interval, and the 2-ball is a closed disk.

neighborhood homeomorphic to $\mathbb{H}^k$, and it is denoted as bd $\mathbb{X}$. The boundary is always either empty or a $(k-1)$-manifold (without boundary). Why is that true? Note the slight awkwardness of language: a manifold with boundary is in general not a manifold, but a manifold is always a manifold with boundary, namely with an empty boundary. An example of a k-manifold with a (nonempty) boundary is the k-ball; see Figure 3.16. Its boundary is the $(k-1)$-sphere, bd $\mathbb{B}^k = \mathbb{S}^{k-1}$.

Orientability

Manifolds with or without boundary can be either orientable or nonorientable. The distinction is a global property that cannot be observed locally. Intuitively, we can imagine a $(k+1)$-dimensional ant walking on the k-manifold. At any moment, the ant is on one side of the local neighborhood with which it is in contact. The manifold is *nonorientable* if there is a walk that brings the ant back to the same neighborhood but now on the other side, and it is *orientable* if no such path exists. Examples of nonorientable manifolds, one with and one without a boundary, are the Möbius strip and the Klein bottle, both illustrated in Figure 3.17.

Imagine the boundary of a solid shape in our everyday three-dimensional space. This boundary is a 2-manifold, and it bounds the interior of the shape on one side and the exterior on the other. The 2-manifold must therefore be

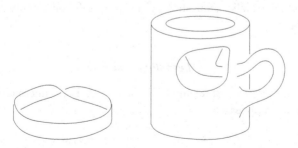

Figure 3.17. The Möbius strip to the left is bounded by a single circle. The Klein mug to the right is drawn with a cutaway view to show a piece of the handle after it passes through the surface of the mug.

orientable. At it turns out, every 2-manifold embedded in $\mathbb{R}^3$ separates inside from outside and is therefore orientable. The Klein bottle is nonorientable and can therefore not exist in $\mathbb{R}^3$. More precisely, the Klein bottle has no *embedding* in $\mathbb{R}^3$; that is, there is no map from the Klein bottle to $\mathbb{R}^3$ whose restriction to the image is a homeomorphism. Any attempt to embed it produces self-intersections, such as the handle that passes through the side of the mug in Figure 3.17. On the other hand, there are obviously 2-manifolds with boundary that can be embedded in $\mathbb{R}^3$, and the Möbius strip is one example.

Bibliographic notes

Point set topology or general topology is an old and well-established branch of mathematics. A classic text on the topic is the book by John Kelley [3]. Manifolds are studied primarily in the context of differential structures. The topological aspects of such structures are emphasized in the text by Guillemin and Pollack [2]. The difference between orientable and nonorientable 2-manifolds is discussed in some length in the popular novel about life in a flat land by Edwin Abbott [1]. The same issue for 3-manifolds is addressed from a more mathematical viewpoint in the book by Jeffrey Weeks [4].

[1] E. A. Abbott. *Flatland*. Sixth edition, Dover, New York, 1952.
[2] V. Guillemin and A. Pollack. *Differential Topology*. Prentice Hall, Englewood Cliffs, New Jersey, 1974.
[3] J. L. Kelley. *General Topology*. Springer-Verlag, New York, 1955.
[4] J. R. Weeks. *The Shape of Space*. Marcel Dekker, New York, 1985.

3.4 Euler characteristic

A topological invariant that predated the creation of topology as a field within mathematics is the Euler characteristic of a space. This section introduces the Euler characteristic, talks about shelling, and proves the shellability of triangulations of the disk.

Alternating sums

The *Euler characteristic* of a simplicial complex K is the alternating sum of the number of simplices,

$$\chi(K) = s_0 - s_1 + s_2 - \cdots + (-1)^d s_d,$$

where $d = \dim K$ and s_i is the number of i-simplices in K. It is common to omit the (-1)-simplex from the sum. A simplex of even dimension contributes 1 and a simplex of odd dimension contributes -1; therefore

$$\chi(K) = \sum_{\emptyset \neq \sigma \in K} (-1)^{\dim \sigma}.$$

As an example consider the complex $B = \mathrm{Cl}\,\sigma$, which consists of σ and all its faces. Assuming $d = \dim \sigma$, we have

$$\chi(B) = \sum_{i=0}^{d} (-1)^i \binom{d+1}{i+1}$$

$$= (1-1)^{d+1} - (-1)^{-1} \binom{d+1}{0}$$

$$= 1.$$

The boundary complex of B is $S = B - \{\sigma\}$, and its Euler characteristic is

$$\chi(S) = \chi(B) - (-1)^{\dim \sigma}$$

$$= 1 - (-1)^{\dim \sigma}.$$

This is zero if the dimension of σ is even and two if the dimension is odd.

Shelling

Consider the case in which K is a triangulation of the closed disk, $\mathbb{B}^2$. We draw K in the plane and get a triangulated polygon like the one shown in Figure 3.18. Our goal is to prove $\chi(K) = 1$. Since K can be any triangulation of $\mathbb{B}^2$, this amounts to proving that *every* triangulation of the closed disk has Euler characteristic equal to one. We use the concept of shelling to prove this claim.

A *shelling* of K is an ordering of the triangles such that each prefix defines a triangulation of $\mathbb{B}^2$. K is *shellable* if it has a shelling. A shelling constructs

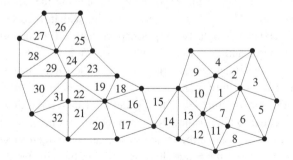

Figure 3.18. The numbers specify a shelling of the triangulation.

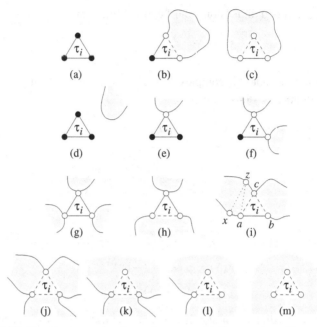

Figure 3.19. The 13 ways a triangle can intersect with the complex of its predecessors. Only cases (a), (b), and (c) occur in a shelling.

K without every creating pinch points, disconnected pieces, or holes along the way. Suppose $\tau_1, \tau_2, \ldots, \tau_n$ is a shelling. When we add τ_i, for $i \geq 2$, its intersection with the complex of its predecessors either consists of a single edge together with its two vertices, or of two edges together with their three vertices; see cases (b) and (c) in Figure 3.19. Case (a) occurs only for τ_1. Cases (d) to (m) cannot occur in a shelling because the union of either the first $i - 1$ or the first i triangles is not homeomorphic to $\mathbb{B}^2$.

Let K_i be the closure of the set of first i triangles in the shelling. At the beginning we have $\chi(K_1) = 1$. If τ_i shares one edge with its predecessors, then we effectively add one triangle, two edges, and one vertex. If τ_i shares two edges, we add one triangle and one edge. In either case we have $\chi(K_i) = \chi(K_{i-1})$. In other words, the Euler characteristic remains the same during the entire construction, and therefore $\chi(K) = \chi(K_n) = 1$. Modulo the existence of a shelling we thus proved that *every* triangulation of the closed disk has Euler characteristic equal to one. Although we cannot prove it here, we state that if K and L are triangulations of the same topological space, then $\chi(K) = \chi(L)$. This result is the reason the Euler characteristic can be considered a property of the topological space, rather than of a triangulation of that space.

Shellability

To complete the proof that $\chi(K) = 1$ for every triangulation K of the closed disk, we still need to find a shelling.

Disk Shelling Lemma. Every triangulation of $\mathbb{B}^2$ is shellable.

Proof. We construct a shelling backwards. Specifically, we prove that every triangulation K of $n \geq 2$ triangles has a triangle τ_n that meets the rest in one of the two allowable configurations shown as cases (b) and (c) in Figure 3.19.

Case 1. K contains no interior vertex. We consider the dual graph, whose nodes are the triangles in K and whose arcs connect two nodes if they share a common edge. The dual graph is a tree and therefore contains at least two leaves. A leaf triangle is connected to only one other node, which it meets in one edge as shown in case (b) of Figure 3.19.

Case 2. K contains no leaf. Then K contains at least one interior vertex. Let ab be an edge on the boundary and abc the triangle next to ab. Either abc intersects the rest of K as in case (c) of Figure 3.19, or c is also a boundary vertex, as in case (i) of Figure 3.19. Assuming the latter, we consider the boundary path from a to c that does not pass through b. Let ax and zc be the first and last edges on that path. We are done if one of the two triangles next to ax and zc satisfies the requirements for τ_n. Otherwise, both triangles meet the rest as in case (i), and one of these two triangles shares ac with abc. We continue the search by substituting that triangle for abc. The search cannot continue indefinitely because K has only finite size. Eventually, we find a triangle τ_n that meets the rest in two edges as in case (c) of Figure 3.19.

By induction we may assume that the complex defined by the remaining $n - 1$ triangles has a shelling. We append τ_n and have a shelling of K. □

Cell complexes

The Euler characteristic in not restricted to simplicial complexes and can be computed as the alternating sum of the number of cells in more general complexes. If two complexes have homeomorphic underlying spaces, we require them to have the same Euler characteristic. For a large class of complexes this is true. Without exploring the most general setting that satisfies this requirement, we consider finite *cell complexes*. Its cells are pairwise disjoint open balls, the boundary of a cell is the closure minus the cell, and the boundary of each cell

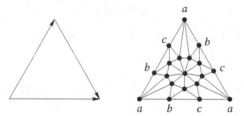

Figure 3.20. The dunce cap to the left consists of one 2-cell, one edge, and one vertex. Its triangulation to the right consists of 27 triangles, 39 edges, and 13 vertices.

is the union of other cells in the complex. An example is the dunce cap constructed from a triangular piece of paper by gluing the three edges, as indicated in Figure 3.20.

Let Z be a finite cell complex. For technical reasons we assume there is a simplicial complex K and a homeomorphism $f : |Z| \to |K|$ such that the image of the closure of each cell is the underlying space of a subcomplex of K. K triangulates $|Z|$ as well as the closure of every cell in $z \in Z$. Let $S \subseteq B \subseteq K$ be subcomplexes that triangulate the boundary and the closure of z. We think of $B - S$ as a triangulation of the interior, which is the cell itself. Since gluing along the boundary does not affect the interior, we can pretend that S triangulates a sphere. The Euler characteristic of $B - S$ is therefore

$$\chi(B - S) = \chi(B) - \chi(S)$$
$$= 1 - [1 - (-1)^d]$$
$$= (-1)^d,$$

where d is the dimension of z. We see that the contribution of $B - S$ to $\chi(K)$ is the same as the contribution of z to $\chi(Z)$. It follows that the Euler characteristics are the same, $\chi(Z) = \chi(K)$. This observation greatly simplifies the hand computation of the Euler characteristic for many spaces.

2-manifolds

A two-dimensional manifold can be constructed from a piece of paper by gluing edges along its boundary. As an example consider the torus, $\mathbb{T}$, which can be constructed from a square by gluing edges in opposite pairs as shown in Figure 3.21. The square, together with its two edges and one vertex, forms a cell complex for the torus, with Euler characteristic

$$\chi(\mathbb{T}) = 1 - 2 + 1 = 0.$$

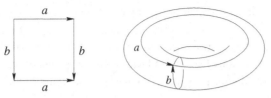

Figure 3.21. Edges with the same label are glued so their arrows agree. After gluing we have two edges and one vertex.

The straightforward treatment of the torus can be extended to general 2-manifolds by using the complete characterization of 2-manifolds, which was one of the major achievements in nineteenth-century mathematics. The list of orientable 2-manifolds consists of the 2-sphere, the torus with one hole, the torus with two holes, and so on. The number of holes is the *genus* of the 2-manifold. The torus with g holes can be constructed from its polygonal schema, which is a regular $4g$-gon with edges

$$a_1 a_2 a_1^- a_2^- a_3 a_4 a_3^- a_4^- \cdots a_{2g-1} a_{2g} a_{2g-1}^- a_{2g}^-,$$

where an edge without minus sign is directed in anticlockwise and one with minus is directed in clockwise order around the $4g$-gon; see Figure 3.22. The g-holed torus is constructed by gluing edges in pairs as indicated by the labels. After gluing we are left with $2g$ edges and one vertex. The Euler characteristic is therefore $\chi(\mathbb{T}_g) = 2 - 2g$. Given a triangulated orientable 2-manifold, we can therefore use the Euler characteristic to compute the genus and decide the topological type of the 2-manifold.

Bibliographic notes

The history of the Euler characteristic is described in an entertaining book by Imre Lakatos, who studies progress in mathematics from a philosophical perspective [3]. The final word on the subject is contained in a seminal paper by Henri Poincaré. He proves that the Euler characteristic is equal to the alternating

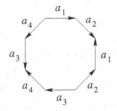

Figure 3.22. The polygonal schema of the double torus.

sum of Betti numbers, which are ranks of homology groups [4]. The Euler characteristic has been developed and applied in a number of directions; see, for example, [5]. An algorithm for constructing a shelling of a triangulated disk is described in a paper by Danaraj and Klee [1]. A treatment of the classification of 2-manifolds and their polygonal schemas can be found in the text by Peter Giblin [2].

[1] G. Danaraj and V. Klee. A representation of 2-dimensional pseudomanifolds and its use in the design of a linear time shelling algorithm. *Ann. Discrete Math.* **2** (1978), 53–63.

[2] P. J. Giblin. *Graphs, Surfaces and Homology.* Second edition, Chapman and Hall, London, 1981.

[3] I. Lakatos. *Proofs and Refutations.* Cambridge Univ. Press, Cambridge, England, 1976.

[4] H. Poincaré. Complément à l'analysis situs. *Rendiconti del Circolo Matematico di Palermo* **13** (1899), 285–343.

[5] Yu. A. Shashkin. *The Euler Characteristic.* Little Math. Library, Mir Publ., Moscow, 1989.

Exercise collection

The credit assignment reflects a subjective assessment of difficulty. A typical question can be answered by using knowledge of the material combined with some thought and analysis.

1. **Union of intervals** (three credits). Let I be a set of n closed intervals on the real line. The union of the intervals is a collection of $m \leq n$ pairwise disjoint intervals. Prove that

$$m = \sum_{i \geq 0} (-1)^i \cdot n_i,$$

   where n_i counts the $(i + 1)$-tuples of intervals with nonempty common intersection.

2. **Regular tetrahedron** (one credit). To prove that the regular tetrahedron does not tile $\mathbb{R}^3$, show that the dihedral angle between two of its triangles is not an integer fraction of 2π. What exactly is that dihedral angle?

3. **Counting simplices** (two credits). Consider the edgewise subdivision of a k-simplex, where each edge is cut into $j \geq 1$ shorter edges. Prove that the number of k-simplices in that subdivision is j^k.

4. **Union of simplices** (two credits). The union of a finite collection of triangles in $\mathbb{R}^2$ is a polyhedron because we can find another set of triangles with the same union whose closure is a simplicial complex. Show that the natural extension of this statement to $\mathbb{R}^d$ is true.

5. **Extendable shelling** (three credits). Let K be a triangulation of the closed disk, let n be the number of triangles, and let $\tau_1, \tau_2, \ldots, \tau_i$ be an ordering of $i < n$ triangles so every prefix defines another triangulation of $\mathbb{B}^2$. For obvious reasons, the sequence from τ_1 to τ_i is called a *partial shelling* of K. Prove that every partial shelling can be extended to a complete shelling,

$$\tau_1, \tau_2, \ldots, \tau_i, \tau_{i+1}, \ldots, \tau_n.$$

6. **Shelling algorithm** (two credits). Let K be a 2-complex that triangulates $\mathbb{B}^2$ with n triangles. Describe an algorithm that constructs a shelling of K in time $O(n)$. You may assume that K is given in a data structure that is convenient for your algorithm.

7. **Classifying manifolds** (two credits). Let K be a triangulation of a 2-manifold.
   (i) Give a linear time algorithm that decides whether or not the 2-manifold is orientable.
   (ii) Assuming it is orientable, give a linear time algorithm that computes the genus of the 2-manifold.

8. **Euler characteristic** (two credits). Compute the Euler characteristic of the following topological spaces:
   (i) the cylinder,
   (ii) the Möbius strip,
   (iii) the Klein bottle, and
   (iv) the solid torus,

9. **Embedding the dunce cap** (three credits). Recall that the dunce cap can be construction from a triangular sheet of paper by gluing all three edges to one. Does the dunce cap have a triangulation that can be geometrically realized in $\mathbb{R}^3$?

4

Surface simplification

This chapter describes an algorithm for simplifying a given triangulated surface. We assume this surface represents a shape in three-dimensional space, and the goal is to represent approximately the same shape with fewer triangles. The particular algorithm combines topological and numerical computations and provides an opportunity to discuss combinatorial topology concepts in an applied situation. Section 4.1 describes the algorithm, which greedily contracts edges until the number of triangles that remain is as small as desired. Section 4.2 studies topological implications and characterizes edge contractions that preserve the topological type of the surface. Section 4.3 interprets the algorithm as constructing a simplicial map and establishes connections between the original and the simplified surfaces. Section 4.4 explains the numerical component of the algorithm used to prioritize edges for contraction.

4.1 Edge contraction algorithm

A triangulated surface is simplified by reducing the number of vertices. This section presents an algorithm that simplifies by repeated edge contraction. We discuss the operation, describe the algorithm, and introduce the error measure that controls which edges are contracted and in what sequence.

Edge contraction

Let K be a 2-complex, and assume for the moment that $|K|$ is a 2-manifold. The contraction of an edge $ab \in K$ removes ab together with the two triangles abx, aby, and it mends the hole by gluing xa to xb and ya to yb, as shown in Figure 4.1. Vertices a and b are glued to form a new vertex c. All simplices in the star of c are new, and the rest of the complex stays the same. To express this more formally, we define the *cone* from a point x to a simplex τ as the union of

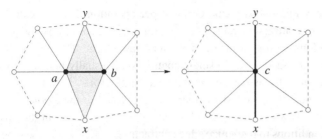

Figure 4.1. The contraction of edge *ab*. Vertices *a* and *b* are glued to a new vertex *c*.

line segments connecting x to points $p \in \tau$, $x \cdot \tau = \text{conv}(\{x\} \cup \tau)$. It is defined only if x is not an affine combination of the vertices of τ. With this restriction, $x \cdot \tau$ is a simplex of one higher dimension, $\dim(x \cdot \tau) = 1 + \dim \tau$. For a *set* of simplices, the cone is defined if it is defined for each simplex, and in this case $x \cdot T = \{x \cdot \tau \mid \tau \in T\}$. We also need generalizations of the star and the link from a single simplex to a set of simplices. Denote the closure without the (-1)-simplex as $\bar{T} = \text{Cl}\, T - \{\emptyset\}$. The *star* and *link* of T are

$$\text{St}\, T = \{\sigma \in K \mid \sigma \geq \tau \in T\},$$

$$\text{Lk}\, T = \text{Cl}\,\text{St}\, T - \text{St}\,\bar{T}.$$

For closed sets T, the link is simply the boundary of the closed star. For example, in Figure 4.1 the link of the set $\overline{ab} = \{ab, a, b\}$ is the cycle of dashed edges and hollow vertices bounding the closed star of $\overline{ab}$. The *contraction* of the edge *ab* is the operation that changes K to $L = K - \text{St}\,\overline{ab} \cup c \cdot \text{Lk}\,\overline{ab}$. This definition applies generally and does not assume that K is a manifold.

Algorithm

The surface represented by K is simplified by performing a sequence of edge contractions. To get a meaningful result, we prioritize the contractions by the numerical error they introduce. Contractions that change the topological type of the surface are rejected. Initially, all edges are evaluated and stored in a priority queue. The process continues until the number of vertices shrinks to the target number m. Let $n \geq m$ be the number of vertices in K.

```
while  n > m and  priority queue non-empty do
      remove top edge ab from priority queue;
      if  contracting ab preserves topology then
            contract ab; n--
      endif
endwhile.
```

The priority queue takes time $O(\log n)$ per operation. Besides extracting the edge whose contraction causes the minimum error, we remove edges that no longer belong to the surface and we add new edges. The number of edges removed and added during a single contraction is usually bounded by a small constant, but in the worst case it can be as large as $n - 1$. Before performing an edge contraction, we test whether or not it preserves the topological type of the surface. This is done by checking all edges and vertices in the link of $\overline{ab}$. Precise conditions to recognize edge contractions that preserve the type will be discussed in Section 4.2.

Hierarchy

We visualize the actions of the algorithm by drawing the vertices as the nodes of an upside-down forest. The contraction of the edge ab maps vertices a and b to a new vertex c. In the forest this is reflected by introducing c as a new node and declaring it the parent of a and b. The leaves of the forest are the vertices of K, and the roots are the vertices of the simplified complex L. The forest is illustrated in Figure 4.2. We define a function $g : \operatorname{Vert} K \to \operatorname{Vert} L$ that maps each vertex $a \in K$ to the root $u = g(a)$ of the tree in which a is a leaf. The preimage of $u \in L$ is the set of leaves $g^{-1}(u) \subseteq K$ of the tree with root u. The preimages of the roots partition the set of leaves,

$$\operatorname{Vert} K = \bigcup_{u \in L} g^{-1}(u).$$

Let ab be an edge in K and set $u = g(a), v = g(b)$. If $u \neq v$ then ab still exists in L, or rather it corresponds to an edge, namely to $uv \in L$. Else, ab contracts to vertex $u = v \in L$. Similarly, a triangle $abd \in K$ corresponds to $uvw \in L$ if $u = g(a), v = g(b), w = g(d)$ are pairwise different. If two of u, v, w are the same, then abd contracts to an edge, and if $u = v = w$, then abd contracts to this vertex in L.

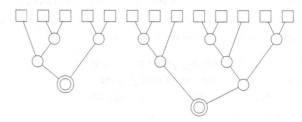

Figure 4.2. Vertices of K are shown as square nodes, intermediate vertices as circle nodes, and vertices of the final complex L as double circle nodes.

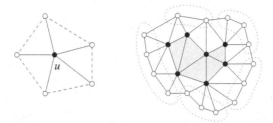

Figure 4.3. Vertex u and its star to the left and the corresponding piece of K to the right. The solid vertices on the right are preimages of u and the hollow vertices are preimages of the neighbors of u.

Numerical error

As mentioned above, a vertex $u \in$ Vert L represents a subset $g^{-1}(u) \subseteq$ Vert K. It makes sense to measure the numerical error at u by comparing u to the part of the original surface it represents. Specifically, we define the error as the sum of square distances of u from the planes spanned by triangles in the star of $g^{-1}(u)$. Figure 4.3 illustrates this idea by showing a vertex $u \in L$ and the triangles in the star of $g^{-1}(u)$. The preimage of u is the collection of seven solid vertices in the right half of the figure. The star of the preimage contains the five shaded triangles and the ring of white triangles around them. The shaded triangles have all their vertices in $g^{-1}(u)$, and the white triangles have either one or two vertices in the preimage.

Let H_u be the set of planes spanned by triangles in St $g^{-1}(u)$. The sum of square distances is defined for every point in $\mathbb{R}^3$, so we can think of the error measure as a function $E_u : \mathbb{R}^3 \to \mathbb{R}$. This function has a unique minimum, unless the normals to the planes in H_u span less than $\mathbb{R}^3$. We can therefore choose u at the point in space where E_u attains its minimum. If the linear subspace spanned by the normals is two-dimensional, then there is a line of minima, and if it is one dimensional, then there is a plane of minima. In both cases we add constraints to pin down u.

Inclusion-exclusion

We will see in Section 4.4 that E_u can be represented by a single symmetric four-by-four matrix $\mathbf{Q}_u$, no matter how many planes there are in H_u. Define $H_w = H_u \cup H_v$. We have $E_w = E_u + E_v - E_{uv}$, where $E_{uv} : \mathbb{R}^3 \to \mathbb{R}$ maps a point in $\mathbb{R}^3$ to the sum of square distances from planes in $H_{uv} = H_u \cap H_v$. We also have

$$\mathbf{Q}_w = \mathbf{Q}_u + \mathbf{Q}_v - \mathbf{Q}_{uv}.$$

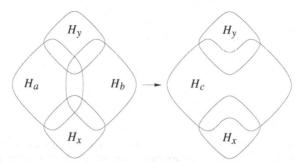

Figure 4.4. Effect of edge contraction on sets of planes used in computing the error.

It is generally not possible to construct $\mathbf{Q}_{uv}$ directly from $\mathbf{Q}_u$ and $\mathbf{Q}_v$. Constructing H_{uv} and computing $\mathbf{Q}_{uv}$ from this set can be expensive. Instead, we maintain matrices $\mathbf{Q}$ for all vertices, edges, and triangles such that $\mathbf{Q}_{ab} = \mathbf{Q}_a \cap \mathbf{Q}_b$ and $\mathbf{Q}_{abd} = \mathbf{Q}_a \cap \mathbf{Q}_b \cap \mathbf{Q}_d$ for all edges ab and all triangles abd. We revisit the contraction of the edge ab. The error function of the new vertex c is given by the matrix $\mathbf{Q}_c = \mathbf{Q}_a + \mathbf{Q}_b - \mathbf{Q}_{ab}$. For every vertex $x \in \mathrm{Lk}\,ab$ there is a new edge xc with error function represented by $\mathbf{Q}_{xc} = \mathbf{Q}_{xa} + \mathbf{Q}_{xb} - \mathbf{Q}_{xab}$, as illustrated in Figure 4.4. We will see later that the matrices for edges are not just useful for correctly computing the matrices for vertices; they also represent meaningful geometric information about edges and their relation to the original surface triangulation.

Bibliographic notes

The problem of simplifying a triangulated surface has its origin in computer graphics, where rendering speed depends on the number of triangles used to represent a shape. The idea of using edge contractions for surface simplification appeared first in a paper by Hoppe et al. [5]. Edge contractions are used together with other local surface modification operations in an attempt to optimize a measure of distance between the original and the simplified surface. Hoppe [3] revisits the idea and shows how to use a given sequence of contractions for efficiently adjusting the level of detail of the surface representation. The idea of measuring error as the sum of square distances from gradually accumulating planes is from Garland and Heckbert [1]. The good quality of the resulting simplifications has intensified the experimental research and has led to variations, such as error measures that account for color and texture of triangulated surfaces [2, 4].

[1] M. Garland and P. S. Heckbert. Surface simplification using quadratic error metrics. *Comput. Graphics*, Proc. SIGGRAPH, 1997, 209–216.

[2] M. Garland and P. S. Heckbert. Simplifying surfaces with color and texture using quadratic error metrics. In "Proc. Visualization, 1998," IEEE Computer Society Press, Los Alamitos, California, 1998, 279–286.

[3] H. Hoppe. Progressive meshes. *Comput. Graphics*, Proc. SIGGRAPH, 1996, 99–108.

[4] H. Hoppe. New quadric metric for simplifying meshes with appearance attributes. In "Proc. Visualization, 1999," IEEE Computer Society Press, Los Alamitos, California, 1999.

[5] H. Hoppe, T. DeRose, T. Duchamp, J. McDonald, and W. Stützle. Mesh optimization. *Comput. Graphics*, Proc. SIGGRAPH, 1993, 19–26.

4.2 Preserving topology

The surface simplification algorithm of the last section works by contracting edges. We preserve the topological type by rejecting the contraction of edges that would change it. This section describes local conditions that characterize type-preserving edge contractions. We first study manifolds, then manifolds with boundary, and finally general 2-complexes.

Manifolds

Suppose K is a 2-complex that triangulates a 2-manifold. Then every point $x \in |K|$ has a neighborhood homeomorphic to an open disk. To avoid lengthy sentences, we just say the neighborhood *is* an open disk. This implies that in particular the star of every vertex u is an open disk. Strictly speaking, this statement makes sense only if we replace the star by its underlying space, which we define as the union of simplex interiors, which is the set difference between the underlying spaces of two complexes:

$$|\operatorname{St} u| = \bigcup_{\tau \in \operatorname{St} u} \operatorname{int} \tau$$
$$= |\operatorname{Cl} \operatorname{St} u| - |\operatorname{Cl} \operatorname{St} u - \operatorname{St} u|.$$

The condition on vertex stars is also sufficient. In other words, $|K|$ is a 2-manifold if and only if $|\operatorname{St} K| \approx \mathbb{R}^2$ for every vertex $u \in K$.

Now consider the contraction of an edge ab of K. Whether or not the contraction preserves the topological type depends on how the links of a and b meet. On a 2-manifold, the link of each vertex is a circle. In Figure 4.5, to the left, the two circles intersect in two points and the contraction preserves the topological type. To the right, the circles intersect in a point and an edge, and in this case the contraction pinches the manifold along a newly formed edge, which forms the base of a fin similar to the one in Figure 4.9.

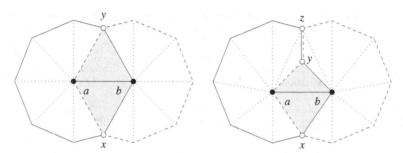

Figure 4.5. The edges of the link of a are solid, and those of the link of b are dashed.

Link condition

The condition that distinguishes topology-preserving edge contractions from others is that the vertex links intersect in the link of the edge.

Link Condition Lemma A. Let K be the triangulation of a 2-manifold. The contraction of $ab \in K$ preserves the topological type if and only if $\mathrm{Lk}\, a \cap \mathrm{Lk}\, b = \mathrm{Lk}\, ab$.

Proof. We prove only the more difficult direction, which is from the link condition to $|K| \approx |L|$. Since $|K|$ is a 2-manifold, the link of ab consists of exactly two vertices x and y, as shown in the left picture of Figure 4.5. The links of a and of b are two circles that meet at x and y. The outer pieces of the two circles glued at x and y form another circle, which is the link of $\overline{ab}$ in K and also the link of c in L. We construct isomorphic subdivisions of K and L by mapping the common link to the boundary of a regular n-gon in the plane, as shown in Figure 4.6. The stars of $\overline{ab}$ and of c are mapped to two triangulations of the n-gon. We

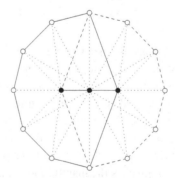

Figure 4.6. The superposition of the images of the stars of $\overline{ab}$ in K and c in L.

superimpose the triangulations and get a decomposition into convex polygonal regions, which we further refine to another triangulation. This triangulation is mapped back to form subdivisions of the stars of $\overline{ab}$ and of c. The link has not been changed, so we can combine the subdivided star of $\overline{ab}$ with the unsubdivided rest of K, and similarly for the star of c and L. The resulting subdivisions of K and L are isomorphic. We can now map corresponding triangles to each other and thus obtain a homeomorphism between $|K|$ and $|L|$. □

A more formal description of how to create the homeomorphism from the isomorphic subdivisions requires simplicial maps, which will be introduced in Section 4.3.

Manifolds with boundary

A triangulation K of a manifold with a nonempty boundary also has vertices whose stars are open half-disks, $|\text{St}\, u| \approx \mathbb{H}^2$. To keep the number of cases small, we add a dummy vertex ω and the cone from ω to each boundary circle. This idea is illustrated in Figure 4.7. The boundary of $|K|$ consists of $\ell \geq 1$ circles triangulated by cycles $C_i \subseteq K$. We fill the holes by adding the cone from ω to every cycle,

$$K^\omega = K \cup \left(\omega \cdot \bigcup_{i=1}^{\ell} C_i \right).$$

In K^ω, every vertex star is an open disk except possibly the star of ω. We denote the link of a vertex u in K^ω as $\text{Lk}^\omega u$. The condition that distinguishes topology-preserving edge contractions from others is now the same as for manifolds.

Link Condition Lemma B. Let K be the triangulation of a 2-manifold with boundary. The contraction of $ab \in K$ preserves the topological type if and only if $\text{Lk}^\omega a \cap \text{Lk}^\omega b = \text{Lk}^\omega ab$.

The proof is only mildly more complicated than that of the weaker Lemma A.

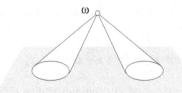

Figure 4.7. The two holes in the manifold are filled by adding the cone from ω to the circles bounding the holes.

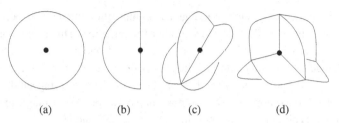

| (a) | (b) | (c) | (d) |

Figure 4.8. The underlying space of the vertex star in (a) is an open disk, in (b) is an open half-disk, in (c) is an open book with four pages, and in (d) is not an open book. The corresponding order of the vertex is zero in (a), one in (b), one in (c), and two in (d).

Open books

To attack the problem for general 2-complexes, we need a better understanding of the different types of neighborhoods that are possible. We classify stars by using a new type of space. The *open book with p pages* is the topological space $\mathbb{K}_p^2$ homeomorphic to the union of p copies of $\mathbb{H}^2$ glued along the common boundary line. For example, the open book with one page is the open half-disk and the open book with two pages is the open disk. The *order* of a simplex $\tau \in K$ is

$$\operatorname{ord} \tau = \begin{cases} 0, & \text{if } |\operatorname{St} \tau| \approx \mathbb{R}^2, \\ 1, & \text{if } |\operatorname{St} \tau| \approx \mathbb{K}_p^2, \ p \neq 2, \\ 2, & \text{otherwise.} \end{cases}$$

Figure 4.8 illustrates the definition with sketches of four different types of vertex stars. The order of an edge in a 2-complex can only be zero or one, and the order of a triangle is always zero.

Boundary

We generalize the notion of boundary in such a way that only triangulations of 2-manifolds have no boundary. At the same time we use the order information to distinguish between different types of boundaries. Specifically, the *jth boundary* of a 2-complex K is the collection of all simplices with order j or higher,

$$\operatorname{Bd}_j K = \{\sigma \in K \mid \operatorname{ord} \sigma \geq j\}.$$

As an example, consider the shark fin complex shown in Figure 4.9. It is constructed by gluing two closed disks along a simple path. This path is a contiguous piece of the boundary of one disk (the fin), and it lies in the interior of the other

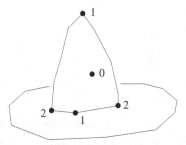

Figure 4.9. The shark fin 2-complex. A few of the vertices are highlighted and marked with their order.

disk. Note that $|K|$ is a 2-manifold if and only if $\mathrm{Bd}_1 K = \mathrm{Bd}_2 K = \emptyset$. The second boundary of a 2-manifold with boundary is empty, but there are other spaces with this property. For example, the sphere together with its equator disk has an empty second boundary. Its first boundary is a circle of edges and vertices (the equator) whose stars are open books of three pages each.

2-complexes

We are now ready to study conditions under which an edge contraction in a general 2-complex preserves the topological type of that complex. As it turns out, there does not exist a local condition that is sufficient and necessary, but there is a characterizing local condition for a more restrictive notion of type preservation. Let L be the 2-complex obtained from K by contracting an edge $ab \in K$. A *local unfolding* is a homeomorphism $f : |K| \to |L|$ that differs from the identity only outside the star of $\overline{ab}$; that is, $f(x) = x$ for all $x \in |K - \mathrm{St}\,\overline{ab}|$. The condition refers to links $\mathrm{Lk}_0^\omega \tau$ in $K^\omega = K \cup (\omega \cdot \mathrm{Bd}_1 K)$ and to links $\mathrm{Lk}_1^\omega \tau$ in $\mathrm{Bd}_1^\omega K = \mathrm{Bd}_1 K \cup (\omega \cdot \mathrm{Bd}_2 K)$.

Link Condition Lemma C. Let K be a 2-complex, ab an edge of K, and L the complex obtained by contracting ab. There is a local unfolding $|K| \to |L|$ if and only if

(i) $\mathrm{Lk}_0^\omega a \cap \mathrm{Lk}_0^\omega b = \mathrm{Lk}_0^\omega ab$ and
(ii) $\mathrm{Lk}_1^\omega a \cap \mathrm{Lk}_0^\omega b = \emptyset$.

The proof that conditions (i) and (ii) suffice for the existence of a local unfolding is similar to proof of sufficiency in Lemma A, only more involved. The necessity of conditions (i) and (ii) requires a somewhat tedious case analysis.

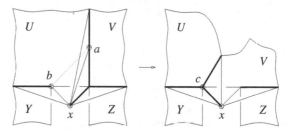

Figure 4.10. The folding chair complex. Each bold edge belongs to three triangles.

Nonlocal homeomorphism

Instead of proving Lemma C, we show that there cannot be a similar condition that recognizes the existence of a general homeomorphism $|K| \to |L|$. The example we use is the folding chair complex displayed in Figure 4.10. Before the contraction of ab, it consists of five triangles in the star of x and four disks U, V, Y, Z glued to the link of x. Vertices a and b belong to the first boundary, but ab does not. It follows that ω violates condition (i) of Lemma C. Hence, there is no local unfolding from $|K|$ to $|L|$. After the contraction there is one less triangle in the star of x; U loses two triangles; and V, Y, Z are unchanged. The contraction of ab exchanges left and right in the asymmetry of the complex. We can find a homeomorphism $|K| \to |L|$ that acts like a mirror and maps U to V, V to U, Y to Z, Z to Y. The homeomorphism is necessarily global. To detect that homeomorphism, we can force any algorithm to look at every triangle of K.

Bibliographic notes

The material of this section is taken from a paper by Dey et al. [1], which studies edge contractions in general simplicial complexes and proves results for 2- and for 3-complexes. The order of a simplex has already been defined in 1960 by Whittlesey [4], although in different words and notation. He uses the concept to study the topological classification of 2-complexes. O'Dunlaing et al. [2] use his results to show that deciding whether or not two 2-complexes have the same topological type is as hard as deciding whether or not two graphs are isomorphic. No polynomial time algorithm is known, but it is also not known whether the graph isomorphism problem is NP-complete [3].

[1] T. K. Dey, H. Edelsbrunner, S. Guha, and D. V. Nekhayev. Topology preserving edge contraction. *Publ. Inst. Math. (Beograd) (N.S.)* **66** (1999), 23–45.
[2] C. O'Dunlaing, C. Watt, and D. Wilkins. Homeomorphism of 2-complexes is equivalent to graph isomorphism. Rept. TCDMATH 98-04, Math. Dept., Trinity College, Ireland, 1998.

[3] M. R. Garey and D. S. Johnson. *Computers and Intractibility. A Guide to the Theory of NP-Completeness.* Freeman, San Francisco, California, 1979.
[4] E. F. Whittlesey. Finite surfaces: a study of finite 2-complexes. *Math. Mag.* **34** (1960), 11–22 and 67–80.

4.3 Simplicial maps

Simplicial maps are piecewise linear (continuous) maps between simplicial complexes. They are used to illuminate the relation between the original surface and the simplified surface generated by the algorithm described in Section 4.1.

Vertex and simplicial maps

We use barycentric coordinates to extend a function between vertex sets to a function between underlying spaces. It is convenient to modify the notation for barycentric coordinates. Each point $x \in |K|$ lies in the interior of exactly one simplex $\sigma \in K$, and we suppose that $p_0, p_1, \ldots, p_k$ are the vertices of σ. Hence $x = \sum_{i=0}^{k} \lambda_i p_i$ with $\sum \lambda_i = 1$ and $\lambda_i > 0$ for all i. For each $u \in$ Vert K we define

$$b_u(x) = \begin{cases} \lambda_i, & \text{if } u = p_i \text{ for } 0 \le i \le k, \\ 0, & \text{if } u \text{ is not a vertex of } \sigma. \end{cases}$$

Instead of fixing x and varying u we now fix u, and vary x. From this point of view we have a map $b_u : |K| \to [0, 1]$, which is zero outside the star of u. It has the shape of a hat that peaks at u, as illustrated in Figure 4.11. In numerical analysis, b_u would be called a base function.

A *vertex map* from K to another simplicial complex L is a function $g :$ Vert $K \to$ Vert L that sends the vertices of a simplex in K to the vertices of a simplex in L. Strictly speaking, g is not a map, but it is called a map because the condition on the images is similar to that required by continuity. The *simplicial*

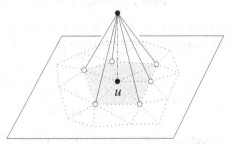

Figure 4.11. We have $b_u(u) = 1$ and $b_u(x) = 0$ for all points x outside the star of u.

map that extends g is $\psi : |K| \to |L|$ defined by

$$\psi(x) = \sum_{u \in \text{Vert } K} b_u(x) \cdot g(u).$$

Indeed, $\psi(u) = g(u)$ for every vertex $u \in K$. The map ψ is continuous because each b_u is continuous.

Simplicial and PL equivalence

A general simplicial map has little predictable structure other than continuity, because the defining vertex map is neither necessarily injective nor necessarily surjective. Even if g is bijective, it is possible that ψ does not reach all points of $|L|$. However, if we assume in addition that its inverse is also a vertex map, then vertices $u_0, u_1, \ldots, u_k$ span a k-simplex $\sigma \in K$ if and only if their images $g(u_0), g(u_1), \ldots, g(u_k)$ span a k-simplex $\tau \in L$. The corresponding simplicial map ψ sends every point $x \in \sigma$ to a unique point $\psi(x) \in \tau$ and vice versa. In other words, $\psi : |K| \to |L|$ is a simplicial homeomorphism. If K and L have a connecting simplicial homeomorphism, then they are abstractly the same complexes. Formally, they are said to be *isomorphic* or *simplicially equivalent*, and this is denoted as $K \cong L$.

We now have an equivalence relation for the class of simplicial complexes. A related equivalence relation for topological spaces requires that they have a common triangulation. If we start with the underlying spaces of two simplicial complexes, we can sometimes generate common triangulations by subdivision. Simplicial complexes K and L are *PL equivalent* if there are isomorphic subdivisions K' of K and L' of L. If K and L are PL equivalent then by definition $|K|$ and $|L|$ are homeomorphic. As it turns out, the other direction is not true in general.

Edge contractions

Suppose K_1 is obtained from $K_0 = K$ by contracting an edge ab. The contraction can be interpreted as a simplicial map $\psi_1 : |K_0| \to |K_1|$, defined by the vertex map

$$g_1(u) = \begin{cases} u, & \text{if } u \notin \{a, b\}, \\ c, & \text{if } u \in \{a, b\}. \end{cases}$$

Both g_1 and ψ_1 are surjective. Another special property of ψ_1 is that the preimage of every point $y \in |K_1|$ is a connected subset of $|K_0|$. More specifically, $\psi^{-1}(y)$

is either a point or a closed line segment in $|K_0|$. This is not true for general edge contractions, but it is for the ones that preserve the topological type of the surface.

The simplification algorithm constructs a sequence of surjective simplicial maps, $\psi_i : |K_{i-1}| \to |K_i|$ for $1 \leq i \leq n - m$. The composition of these maps is

$$\psi = \psi_{n-m} \circ \cdots \circ \psi_2 \circ \psi_1 : |K| \to |L|,$$

where $L = K_{n-m}$. It is the simplicial map extending the vertex map $g = g_{n-m} \circ \cdots \circ g_2 \circ g_1$. Recall that g maps a vertex $u \in K$ to the root of the tree in the hierarchy that contains u as a leaf. The map ψ extends this vertex map to all points of $|K|$. It inherits the property that the preimage of every point in $|L|$ is a connected subset of $|K|$. By continuity, the preimage of every connected subset of $|L|$ is a connected subset of $|K|$. Similarly, the preimage of every open subset in $|L|$ is an open subset of $|K|$. For example, the preimage of the star of $v \in L$ is the star of $g^{-1}(v)$,

$$\psi^{-1}(|\mathrm{St}\, v|) = |\mathrm{St}\, g^{-1}(v)|.$$

Both the underlying space of $\mathrm{St}\, v$ and that of $\mathrm{St}\, g^{-1}(v)$ are connected and open.

Preimages of vertex stars

Consider a collection of $k + 1$ vertices in L. The common intersection of their stars is either empty or the k-simplex spanned by the vertices in the collection. This implies that the nerve of the vertex stars is isomorphic to the complex,

$$\mathrm{Nrv}\, \{\mathrm{St}\, v \mid v \in L\} \cong L.$$

The common intersection of a collection of vertex stars is nonempty if and only if the common intersection of their preimages is nonempty. Hence, also the nerve of stars of the preimages is isomorphic to the complex,

$$\mathrm{Nrv}\, \{\mathrm{St}\, g^{-1}(v) \mid v \in L\} \cong L.$$

In other words, the covering of L by vertex stars corresponds to a covering of K by open sets, which are stars of preimages of vertices. Figure 4.12 illustrates that the sets in this covering form the same overlap pattern as do the vertex stars.

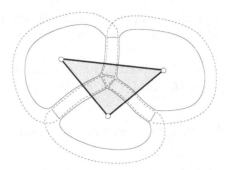

Figure 4.12. The solid curves bound the preimages of the three vertices, and the dashed curves bound the preimages of their stars.

Bibliographic notes

A good background source on simplicial maps is the book on algebraic topology by Munkres [3]. During the early years of combinatorial topology, it was conjectured that two simplicial complexes are PL equivalent if and only if their underlying spaces are homeomorphic. This became known as the Hauptvermutung (German for main conjecture). It was verified for 2-complexes and for 3-complexes; see the book by Moise [2]. In 1961 the conjecture was disproved for general simplicial complexes by John Milnor, who used two seven-dimensional simplicial complexes for the counterexample [1].

[1] J. Milnor. Two complexes which are homeomorphic but combinatorially distinct. *Ann. of Math.* **74** (1961), 575–590.
[2] E. Moise. *Geometric Topology in Dimensions two and three.* Springer-Verlag, New York, 1977.
[3] J. R. Munkres. *Elements of Algebraic Topology.* Addison-Wesley, Redwood City, 1984.

4.4 Error measure

The surface simplification algorithm measures the error of an edge contraction as the sum of square distances of a point from a collection of planes. This section develops the details of this error measure.

Signed distance

A plane with unit normal vector v_i and offset δ_i contains all points p whose orthogonal projection onto the line defined by v_i is $-\delta_i \cdot v_i$,

$$h_i = \{p \in \mathbb{R}^3 \mid p^T \cdot v_i = -\delta_i\},$$

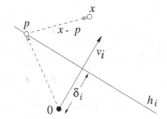

Figure 4.13. We use the unit normal vector to define the signed distance from h_i such that v_i points from the negative to the positive side.

as illustrated in Figure 4.13. The signed distance of a point $x \in \mathbb{R}^3$ from the plane h_i is

$$d(x, h_i) = (x - p)^T \cdot v_i$$
$$= x^T \cdot v_i + \delta_i$$
$$= \mathbf{x}^T \cdot \mathbf{v}_i,$$

where $\mathbf{x}^T = (x^T, 1)$ and $\mathbf{v}_i^T = (v_i^T, \delta_i)$. In other words, the signed distance in $\mathbb{R}^3$ can be expressed as a scalar product in $\mathbb{R}^4$, as illustrated in Figure 4.14.

Fundamental quadric

The sum of square distances of a point x from a collection of planes H is

$$E_H(x) = \sum_{h_i \in H} d^2(x, h_i)$$
$$= \sum_{h_i \in H} (\mathbf{x}^T \cdot \mathbf{v}_i) \cdot (\mathbf{v}_i^T \cdot \mathbf{x})$$
$$= \mathbf{x}^T \cdot \left(\sum_{h_i \in H} \mathbf{v}_i \cdot \mathbf{v}_i^T \right) \cdot \mathbf{x},$$

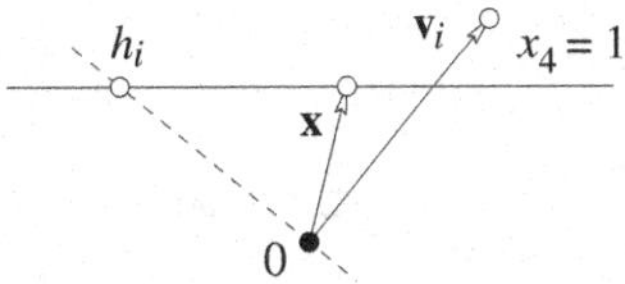

Figure 4.14. The three-dimensional space $x_4 = 1$ is represented by the horizontal line. It contains point x and plane h_i, which in the one-dimensional representation are both points.

where

$$\mathbf{Q} = \sum \mathbf{v}_i \cdot \mathbf{v}_i^T = \begin{pmatrix} A & B & C & D \\ B & E & F & G \\ C & F & H & I \\ D & G & I & J \end{pmatrix}$$

is a symmetric four-by-four matrix referred to as the *fundamental quadric* of the map $E_H : \mathbb{R}^3 \to \mathbb{R}$. The sum of square distances is nonnegative, so $\mathbf{Q}$ is positive semidefinite. The error of an edge contraction is obtained from an error function such as $E = E_H$. Let $\mathbf{x}^T = (x_1, x_2, x_3, 1)$ and note that

$$
\begin{aligned}
E(x) &= \mathbf{x}^T \cdot \mathbf{Q} \cdot \mathbf{x} \\
&= Ax_1^2 + Ex_2^2 + Hx_3^2 \\
&\quad + 2(Bx_1x_2 + Cx_1x_3 + Fx_2x_3) \\
&\quad + 2(Dx_1 + Gx_2 + Ix_3) \\
&\quad + J.
\end{aligned}
$$

We see that E is a quadratic map that is nonnegative and unbounded. Its graph can only be an elliptic paraboloid, as illustrated in Figure 4.15. In other words, the preimage of a constant error value ϵ, $E^{-1}(\epsilon)$, is an ellipsoid. Degenerate ellipsoids are possible, such as cylinders with elliptic cross-sections and pairs of planes.

Error

The *error* of the edge contraction $ab \to c$ is the minimum value of $E(x) = E_H(x)$ over all $x \in \mathbb{R}^3$, where H is the set of planes spanned by triangles in

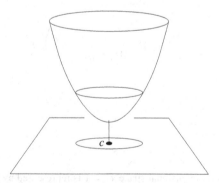

Figure 4.15. Illustration of $E = E_H$ in one lower dimension. The cross-section at a fixed height ϵ is an ellipse.

the preimage of the star of the new vertex c. The geometric location of c is the point x that minimizes E. In the nondegenerate case, this point is unique and can be computed by setting the gradient $\nabla E = (\partial E/\partial x_1, \partial E/\partial x_2, \partial E/\partial x_3)$ to zero. The derivative with respect to x_i is

$$\frac{\partial E}{\partial x_i}(x) = \frac{\partial \mathbf{x}^T}{\partial x_i} \cdot \mathbf{Q} \cdot \mathbf{x} + \mathbf{x}^T \cdot \mathbf{Q} \cdot \frac{\partial \mathbf{x}}{\partial x_i}$$

$$= \mathbf{Q}_i^T \cdot \mathbf{x} + \mathbf{x}^T \cdot \mathbf{Q}_i$$

$$= 2\mathbf{Q}_i^T \cdot \mathbf{x},$$

where $\mathbf{Q}_i^T$ is the ith row of $\mathbf{Q}$. The point $c \in \mathbb{R}^3$ that minimizes $E(x)$ is the solution to the system of three linear equations $Q \cdot x + q = 0$, where

$$Q = \begin{pmatrix} A & B & C \\ B & E & F \\ C & F & H \end{pmatrix}, \qquad q = \begin{pmatrix} D \\ G \\ I \end{pmatrix}.$$

Hence $c = Q^{-1} \cdot (-q)$, and the sum of square distances of c from the planes in H is $E(c)$. The equation for c sheds light on the possible degeneracies. The nondegenerate case corresponds to rank $Q = 3$; the case of an elliptic cylinder corresponds to rank $Q = 2$; and the case of two parallel planes corresponds to rank $Q = 1$. Rank zero is not possible because Q is the nonempty sum of products of unit vectors.

Eigenvalues and eigenvectors

We may translate the planes by $-c$ such that E attains its minimum at the origin. In this case $D = G = I = 0$ and $J = E(0)$. The shape of the ellipsoid $E^{-1}(\epsilon)$ can be described by the eigenvalues and eigenvectors of Q. By definition, the *eigenvectors* are unit vectors x that satisfy $Q \cdot x = \lambda \cdot x$. The value of λ is the corresponding *eigenvalue*. The eigenvalues are the roots of the *characteristic polynomial* of Q, which is

$$P(\lambda) = \det \begin{pmatrix} A - \lambda & B & C \\ B & E - \lambda & F \\ C & F & H - \lambda \end{pmatrix}$$

$$= \det Q - \lambda \cdot \text{dtr } Q + \lambda^2 \cdot \text{tr } Q - \lambda^3,$$

where $\det Q$ is the determinant, $\text{dtr } Q$ is the sum of cofactors of the three diagonal elements, and $\text{tr } Q$ is the trace of Q. For symmetric positive semi-definite matrices, the characteristic polynomial has three nonnegative roots,

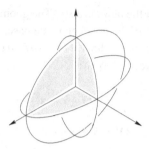

Figure 4.16. The ellipsoid is indicated by drawing the elliptic cross-sections along the three symmetry planes spanned by the eigenvectors.

$\lambda_1 \geq \lambda_2 \geq \lambda_3 \geq 0$. Once we have an eigenvalue, we can compute the corresponding eigenvector to span the null space of the underconstrained system $(Q - \lambda) \cdot x = 0$.

What is the geometric meaning of eigenvectors and eigenvalues? For symmetric matrices, the eigenvectors are pairwise orthogonal, or if there are multiple eigenvalues the eigenvectors can be chosen pairwise orthogonal. They can thus be viewed as defining another coordinate system for $\mathbb{R}^3$. The three symmetry planes of the ellipsoid $E^{-1}(\epsilon)$ coincide with the coordinate planes of this new system; see Figure 4.16. We can write the error function as

$$E(x) = \mathbf{x}^T \cdot \begin{pmatrix} \lambda_1 & & & \\ & \lambda_2 & & \\ & & \lambda_3 & \\ & & & J \end{pmatrix} \cdot \mathbf{x}$$

$$= \lambda_1 x_1^2 + \lambda_2 x_2^2 + \lambda_3 x_3^2 + J.$$

Since $E(x) \geq 0$ for every $x \in \mathbb{R}^3$, this proves that the three eigenvalues are indeed real and nonnegative. The preimage for a fixed error $\epsilon > J$ is the ellipsoid with axes of half-lengths $\sqrt{(\epsilon - J)/\lambda_i}$ for $i = 1, 2, 3$.

Bibliographic notes

The idea of using the sum of square distances from face planes for surface simplification is from Garland and Heckbert [1]. Eigenvalues and eigenvectors of matrices are topics in linear algebra. A very readable introductory text is the book by Gilbert Strang [2].

[1] M. Garland and P. S. Heckbert. Surface simplification using quadratic error metrics. *Computer Graphics*, Proc. SIGGRAPH, 1997, 209–216.
[2] G. Strang. *Introduction to Linear Algebra*. Wellesley-Cambridge Press, Wellesley, Massachusetts, 1993.

Exercise collection

The credit assignment reflects a subjective assessment of difficulty. A typical question can be answered by using knowledge of the material combined with some thought and analysis.

1. **Stars and links** (one credit). Let K be a 2-complex that triangulates the closed disk, $\mathbb{B}^2$. Let a and b be interior vertices; u, v, w boundary vertices; ab, au, uv interior edges; and vw a boundary edge. Draw K such that it contains (among others) five vertices and four edges as specified. Furthermore, draw the star and the link of each of the following subsets of K: $\{a\}$, $\{u\}$, $\{ab\}$, $\{ab, a, b\}$, $\{au, a, u\}$, $\{uv, u, v\}$, $\{vw\}$, and $\{vw, v, w\}$.

2. **Subdivision and nerve** (two credits). Let K be a simplicial complex and Sd K its barycentric subdivision. For each vertex $u \in K$ consider the star of u in Sd K and the closure of that star.
   (i) What is the nerve of the collection of vertex stars?
   (ii) What is the nerve of the collection of closed vertex stars?

3. **Irreducibility** (three credits). A 2-complex K is *irreducible* if the contraction of ab changes the topological type of K for every edge $ab \in K$. Prove that the only irreducible triangulation of the 2-sphere is the boundary complex of the tetrahedron.

4. **Necessity of the link condition** (three credits). Let K be the triangulation of a 2-manifold. Recall the Link Condition Lemma A, which says that the contraction of an edge $ab \in K$ preserves the topological type if and only if $\operatorname{Lk} a \cap \operatorname{Lk} b = \operatorname{Lk} ab$. We proved the sufficiency of the condition in Section 4.3. Prove the necessity of the condition. In other words, prove that $\operatorname{Lk} a \cap \operatorname{Lk} b \neq \operatorname{Lk} ab$ implies that the contraction of ab changes the topological type of the 2-manifold.

5. **Simplicial map** (two credits). Let $\varphi : |K| \to |L|$, $\psi : |L| \to |M|$ be simplicial maps, and suppose they both are injective and have the property that the preimage of a point is either empty or connected. Prove that $\psi \circ \varphi : |K| \to |M|$ has the same property, namely it is injective and the preimage of every point in M is either empty or connected.

6. **Square distance minimization** (three credits). Let S be a finite set of points in $\mathbb{R}^2$. Let $f : \mathbb{R}^2 \to \mathbb{R}$ map each point x in the plane to the sum of square distances,

$$f(x) = \sum_{p \in S} \|x - p\|^2.$$

   (i) Show that f is a quadratic function and that it has a unique minimum.
   (ii) At which point does f attain its minimum?

(iii) Given two disjoint finite sets $S_1, S_2 \subseteq \mathbb{R}^2$ together with their maps f_1, f_2, show how to compute the map f for $S = S_1 \cup S_2$ in constant time.

7. **Points, lines, and planes** (two credits). Let S be a finite set of points in $\mathbb{R}^2$.

   (i) Construct a finite set H of lines such that the sum of square distances to points in S is the same as the sum of square distances to lines in H. Formally,

$$\sum_{p \in S} \|x - p\|^2 = \sum_{h \in H} d(x, h)^2,$$

   where $d(x, h) = \min\{\|x - y\| \mid y \in h\}$.

   (ii) Are the lines solving question (i) unique?

   (iii) Generalize your solution to the case in which S is a finite set of lines and H is a finite set of planes in $\mathbb{R}^3$.

5

Delaunay tetrahedrizations

This chapter extends what we have learned about Delaunay triangulations from two to three dimensions. Almost everything that will be said generalizes readily to four and higher dimensions. It is therefore tempting to introduce a positive integer d and write the entire chapter for the more general d-dimensional case. We resist the temptation in the interest of specificity and focus our attention on the three-dimensional case. Section 5.1 introduces Voronoi diagrams and Delaunay tetrahedrizations and explains their relation to boundary complexes of convex polyhedra in $\mathbb{R}^4$. Section 5.2 generalizes all constructions to points with real weights. Section 5.3 extends the flip operation for Delaunay triangulations to three and higher dimensions by using classic theorems in convex geometry. Section 5.4 describes and analyzes a randomized algorithm that constructs a Delaunay tetrahedrization by adding one point at a time.

5.1 Lifting and polarity

The Delaunay tetrahedrization of a finite set of points in $\mathbb{R}^3$ is dual to the Voronoi diagram of the same set. This section introduces both concepts and shows how they can be obtained as projections of the boundary of convex polyhedra.

Voronoi diagrams

The *Voronoi region* of a point p in a finite collection $S \subseteq \mathbb{R}^3$ is the set of points at least as close to p as to any other point in S:

$$V_p = \{x \in \mathbb{R}^3 \mid \|x - p\| \le \|x - q\|, \ \forall q \in S\}.$$

Each inequality defines a closed half-space, and V_p is the intersection of a finite collection of such half-spaces. In other words, V_p is a convex polyhedron,

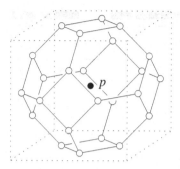

Figure 5.1. The Voronoi polyhedron of a point in a body-centered cube lattice. The relevant neighbors of the cube center p are the eight corners of the cube and the centers of the six adjacent cubes.

maybe like the one shown in Figure 5.1. In the generic case, every vertex of V_p belongs to only three facets and three edges of the polyhedron. If V_p is bounded then it is the convex hull of its vertices. It is also possible that V_p is unbounded. This is the case if and only if there is a plane through p with all points of S on or on one side of the plane.

The Voronoi regions together with their shared facets, edges, and vertices form the *Voronoi diagram* of S. A point x that belongs to k Voronoi regions is equally far from the k generating points. It follows that the k points lie on a common sphere. If the points are in general position, then $k \leq 4$. A Voronoi vertex x belongs to at least four Voronoi regions, and assuming the general position, it belongs to exactly four regions.

Delaunay tetrahedrization

We obtain the *Delaunay tetrahedrization* by taking the dual of the Voronoi diagram. The Delaunay vertices are the points in S. The Delaunay edges connect generators of Voronoi regions that share a common facet. The Delaunay facets connect generators of Voronoi regions that share a common edge. Assuming a general position, each edge is shared by three Voronoi regions and the Delaunay facets are triangles. The Delaunay polyhedra connect generators of Voronoi regions that share a common vertex. Assuming a general position, each vertex is shared by four Voronoi regions and the Delaunay polyhedra are tetrahedra. Consider point p in Figure 5.1. Its Voronoi polyhedron has 14 facets, 36 edges, and 24 vertices. It follows that p belongs to 14 Delaunay edges, 36 Delaunay triangles, and 24 Delaunay tetrahedra, as illustrated in Figure 5.2.

Assuming a general position of the points in S, the Delaunay tetrahedrization is a collection of simplices. To prove that it is a simplicial complex, we still need to show that the simplices avoid improper intersections. We do this by

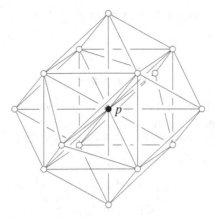

Figure 5.2. The Delaunay neighborhood of a point in a body-centered cube lattice.

introducing geometric transformations that relate Voronoi diagrams and Delaunay tetrahedrizations in $\mathbb{R}^3$ with boundary complexes of convex polyhedra in $\mathbb{R}^4$.

Distance maps

The square distance from $p \in S$ is the map $\pi_p : \mathbb{R}^3 \to \mathbb{R}$ defined by $\pi_p(x) = \|x - p\|^2$. Its graph is a paraboloid of revolution in $\mathbb{R}^4$. We simplify notation by suppressing the difference between a function and its graph. Figure 5.3 illustrates this idea in one lower dimension. Take the collection of all square distance functions defined by points in S. The pointwise minimum is the map $\pi_S : \mathbb{R}^3 \to \mathbb{R}$ defined by

$$\pi_S(x) = \min\{\pi_p(x) \mid p \in S\}.$$

Its graph is the lower envelope of the collection of paraboloids. By definition of the Voronoi region, $\pi_S(x) = \pi_p(x)$ if and only if $x \in V_p$. We can therefore

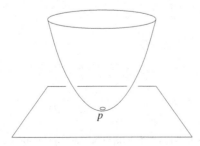

Figure 5.3. The graph of the square distance function of point p in the plane.

think of V_p as the projection of the portion of the lower envelope contributed by the paraboloid π_p.

Linearization

All square distance functions have the same quadratic term, which is $\|x\|^2$. If we subtract that term we get linear functions, namely

$$
\begin{aligned}
f_p(x) &= \pi_p(x) - \|x\|^2 \\
&= (x - p)^T \cdot (x - p) - x^T \cdot x \\
&= -2p^T \cdot x + \|p\|^2.
\end{aligned}
$$

The graph of f_p is a hyperplane in $\mathbb{R}^4$. The same transformation warps the hyperplane $x_4 = 0$ to the upside-down paraboloid Π, defined as the graph of the map defined by $\Pi(x) = -\|x\|^2$. Figure 5.4 shows the result of the transformation applied to the plane and paraboloid in Figure 5.3. We can apply the transformation to the entire collection of paraboloids at once. Each point in $\mathbb{R}^4$ travels vertically, that is, parallel to the x_4-axis. The traveled distance is the square distance to the x_4-axis. Paraboloids go to hyperplanes, intersections of paraboloids go to intersections of hyperplanes, and the lower envelope of the paraboloids goes to the lower envelope of the hyperplanes.

Replace each hyperplane by the closed half-space bounded from above by the hyperplane. The intersection of the half-spaces is a convex polyhedron F in $\mathbb{R}^4$, and the lower envelope of the hyperplanes is the boundary of F. It is a complex of convex faces of dimension 3, 2, 1, 0. Since the transformation moves points vertically, the projection onto $x_4 = 0$ of the lower envelope of paraboloids and the lower envelope of hyperplanes are the same. In particular, the projection of each three-dimensional face of F is a Voronoi region, and the projection of the entire boundary complex is the Voronoi diagram.

Figure 5.4. The plane in Figure 5.3 becomes an upside-down paraboloid, and the paraboloid becomes a plane.

Polarity

We still need to describe what all this has to do with the Delaunay tetrahedrization of S. Instead of addressing this question directly, we first study the relationship between nonvertical hyperplanes and their polar points in $\mathbb{R}^4$.

A nonvertical hyperplane is the graph of a linear function $f : \mathbb{R}^3 \to \mathbb{R}$, which can generally be defined by a point $p \in \mathbb{R}^3$ and a scalar $c \in \mathbb{R}$; that is,

$$f(x) = -2p^T \cdot x + \|p\|^2 - c.$$

The hyperplane parallel to f and tangent to Π is defined by the equation $-2p^T \cdot x + \|p\|^2$. The vertical distance between the two hyperplanes is $|c|$. The *polar point* of f is $g = f^* = (p, -\|p\|^2 + c)$. The vertical distance between g and f is $2|c|$, and the parallel tangent hyperplane lies right in the middle between g and f. Furthermore, the vertical line through g also passes through the point where the tangent hyperplane touches Π. It follows that $g \in \Pi$ if and only if f is tangent to Π. Figure 5.5 shows a few examples of hyperplanes and their polar points in $\mathbb{R}^2$. For nonvertical hyperplanes, the points lying *above, on, below* are unambiguously defined. Let f_1, f_2 be two nonvertical hyperplanes and g_1, g_2 their polar points.

Order Reversal Claim. Point g_1 lies above, on, below hyperplane f_2 if and only if point g_2 lies above, on, below hyperplane f_1.

Proof. Let $g_i = (p_i, -\|p_i\|^2 + c_i)$ for $i = 1, 2$. The algebraic expression for g_1 above f_2 is

$$-\|p_1\|^2 + c_1 > -2p_2^T \cdot p_1 + \|p_2\|^2 - c_2.$$

We move terms left and right and use the fact that scalar products are commutative to get

$$-\|p_2\|^2 + c_2 > -2p_1^T \cdot p_2 + \|p_1\|^2 - c_1.$$

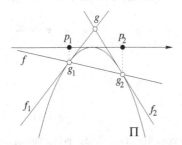

Figure 5.5. Points g_1, g_2, g are polar to the lines (hyperplanes) f_1, f_2, f. Lines f_1, f_2 are warped images of the distance square functions of the points p_1, p_2 on the real line.

This is the algebraic expression for g_2 above f_1. The arguments for point g_1 lying on and below hyperplane f_2 are the same. □

Polar polyhedron

We are now ready to construct the Delaunay tetrahedrization as the projection of the boundary complex of a convex polyhedron in $\mathbb{R}^4$. For each point $p \in S$, let $g_p = (p, -\|p\|^2)$ be the polar point of the corresponding hyperplane. All points g_p lie on the upside-down paraboloid Π, as shown in Figure 5.6. For a nonvertical hyperplane f, we consider the closed half-space bounded from above by f. Let G be the intersection of all such half-spaces that contain all points g_p. G is a convex polyhedron in $\mathbb{R}^4$. Its boundary consists of the upper portion of the convex hull boundary plus the silhouette extended to infinity in the $-x_4$ direction. The Order Reversal Claim implies the following correspondence between G and F. A hyperplane *supports* G if it has nonempty intersection with the boundary and empty intersection with the interior.

Support Claim. A hyperplane f supports G if and only if the polar point $g = f^*$ lies in the boundary of F.

Imagine exploring G by rolling the supporting hyperplane along its boundary. The dual image of this picture is the polar point moving inside the boundary of F. For each k-dimensional face of G we get a $(3-k)$-dimensional face of F and vice versa. An exception is the set of vertical faces of G, which do not correspond to any faces of F, except possibly to faces stipulated at infinity. The relationship between the two boundary complexes is the same as that between the Delaunay tetrahedrization and the Voronoi diagram. The isomorphism between the boundary complex of F and the Voronoi diagram implies the isomorphism between the boundary complex of G (excluding vertical faces) and the Delaunay tetrahedrization. Since the vertices of G project onto points in S, it follows that the boundary complex of G projects onto the Delaunay

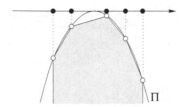

Figure 5.6. The boundary complex of the shaded polyhedron projects onto the Delaunay tetrahedrization of the set of solid points.

tetrahedrization of S. This finally implies that there are no improper intersections between Delaunay simplices. The Delaunay tetrahedrization of a set S of finitely many points in general position is indeed a simplicial complex.

Bibliographic notes

Voronoi diagrams and Delaunay triangulation are named after Georges Voronoi [3] and Boris Delaunay (also Delone) [1]. The concepts themselves are older and can be traced back to prominent mathematicians of earlier centuries, including Friedrich Gauß and René Descartes. The connection to convex polytopes has also been known for a long time. The combinatorial theory of convex polytopes is a well-developed field within mathematics. We refer to the texts by Branko Grünbaum [2] and by Günter Ziegler [4] for excellent sources of the accumulated knowledge in that subject.

[1] B. Delaunay. Sur la sphère vide. *Izv. Akad. Nauk SSSR, Otdelenie Matematicheskii i Estestvennyka Nauk* **7** (1934), 793–800.
[2] B. Grünbaum. *Convex Polytopes*. John Wiley and Sons, London, 1967.
[3] G. Voronoi. Nouvelles applications des paramètres continus à la théorie des formes quadratiques. *J. Reine Angew. Math.* **133** (1907), 97–178, and **134** (1908), 198–287.
[4] G. M. Ziegler. *Lectures on Polytopes*. Springer-Verlag, New York, 1995.

5.2 Weighted distance

The correspondence between Voronoi diagrams and convex polyhedra hints at a generalization of Voronoi and Delaunay diagrams forming a richer class of objects. This section describes this generalization by using points with real weights. Within this larger class of diagrams, we find a symmetry between Voronoi and Delaunay diagrams absent in the smaller class of unweighted diagrams.

Commuting diagram

Figure 5.7 illustrates the correspondence between Voronoi diagrams and Delaunay tetrahedrizations in $\mathbb{R}^3$ and convex polyhedra in $\mathbb{R}^4$, as worked out in Section 5.1. V and D are dual to each other; F is obtained from V through linearization of distance functions; and V is formed by the projections of the boundary complex of F. F and G are polar to each other; G is the convex hull of the points projected onto Π (extended to infinity along the $-x_4$-direction); and D is the projection of the boundary complex of G.

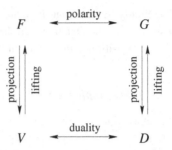

Figure 5.7. Relationship between Voronoi diagram, V, Delaunay tetrahedrization, D, and convex polyhedra, F and G.

We call G an *inscribed* polyhedron because each vertex lies on the upside-down paraboloid Π. Similarly, we call F a *circumscribed* polyhedron because each hyperplane spanned by a 3-face is tangent to Π. Being inscribed or circumscribed is a rather special property. We use weights to generalize the concepts of Voronoi diagrams and Delaunay tetrahedrizations in a way that effectively frees the polyhedra from being inscribed or circumscribed. For technical reasons, we still require that every vertical line intersects F in a half-line and G either in a half-line or the empty set. This is an insubstantial although sometimes inconvenient restriction.

Weighted points

We prepare the definition of weighted Delaunay tetrahedrization by introducing points with real weights. It is convenient to write the weight of a point as the square of a nonnegative real or a nonnegative multiple of the imaginary unit. We think of the weighted point $\hat{p} = (p, P^2) \in \mathbb{R}^3 \times \mathbb{R}$ as the sphere with center $p \in \mathbb{R}^3$ and radius P. The *power* or *weighted distance function* of $\hat{p}$ is the map $\pi_{\hat{p}} : \mathbb{R}^3 \to \mathbb{R}$ defined by

$$\pi_{\hat{p}}(x) = \|x - p\|^2 - P^2.$$

It is positive for points x outside the sphere, zero for points on the sphere, and negative for points inside the sphere. The various cases permit intuitive geometric interpretations of weighted distance. For example, for positive P^2 and x outside the sphere, it is the square length of a tangent line segment connecting x with a point on the sphere. This is illustrated in Figure 5.8. What is it if x lies inside the sphere? In Section 1.1, we saw that the set of points with equal weighted distance from two circles is a line. Similarly, the set of points with equal weighted distance from two spheres in $\mathbb{R}^3$ is a plane. If the

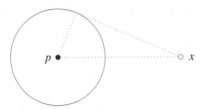

Figure 5.8. The segment px, the tangent segment from x to the circle, and the connecting radius form a right-angled triangle.

two spheres intersect then the plane passes through the intersection circle, and if the two spheres are disjoint and lie side by side then the plane separates the two spheres.

Orthogonality

Given two spheres or weighted points $\hat{p} = (p, P^2)$ and $\hat{q} = (q, Q^2)$, we generalize weighted distance to the symmetric form:

$$\pi_{\hat{p},\hat{q}} = \|p - q\|^2 - P^2 - Q^2.$$

For $Q^2 = 0$, this is the weighted distance from q to $\hat{p}$, and for $P^2 = 0$, this is the weighted distance from p to $\hat{q}$. We call $\hat{p}$ and $\hat{q}$ *orthogonal* if $\pi_{\hat{p},\hat{q}} = 0$. Indeed, if $P^2, Q^2 > 0$ then $\pi_{\hat{p},\hat{q}} = 0$ if and only if the two spheres meet in a circle and the two tangent planes at every point of this circle form a right angle. Orthogonality is the key concept in generalizing Delaunay to weighted Delaunay tetrahedrizations. We call $\hat{p}$ and $\hat{q}$ *further than orthogonal* if $\pi_{\hat{p},\hat{q}} > 0$.

Let us contemplate for a brief moment how weights affect the lifting process. The graph of the weighted distance function is a paraboloid whose zero set, $\pi_{\hat{p}}^{-1}(0)$, is the sphere $\hat{p}$. We can linearize as before and get a hyperplane defined by

$$\begin{aligned} f_{\hat{p}}(x) &= \pi_{\hat{p}}(x) - \|x\|^2 \\ &= -2p^T \cdot x + \|p\|^2 - P^2. \end{aligned}$$

We can also polarize and get

$$g_{\hat{p}} = (p, -\|p\|^2 + P^2).$$

Orthogonality between two spheres now translates to a point–hyperplane incidence.

Orthogonality Claim. Spheres $\hat{p}$ and $\hat{q}$ are orthogonal if and only if point $g_{\hat{p}}$ lies on the hyperplane $f_{\hat{q}}$.

Proof. The algebraic expression for $g_{\hat{p}} \in f_{\hat{q}}$ is

$$-2q^T \cdot p + \|q\|^2 - Q^2 = -\|p\|^2 + P^2.$$

This is equivalent to

$$(p - q)^T \cdot (p - q) - P^2 - Q^2 = 0,$$

which is equivalent to $\pi_{\hat{p},\hat{q}} = 0$. $\square$

Weighted Delaunay tetrahedrization

Let S be a finite set of spheres. Depending on the application, we think of an element of S as a point in $\mathbb{R}^3$ or a weighted point in $\mathbb{R}^3 \times \mathbb{R}$. The weighted distance can be used to construct the *weighted Voronoi diagram*, and the *weighted Delaunay tetrahedrization* is dual to that diagram, as usual. Instead of going through the technical formalism of the construction, which is pretty much the same as for unweighted points, we illustrate the concept in Figure 5.9. For unweighted points, a tetrahedron belongs to the Delaunay tetrahedrization if and only if the circumsphere passing through the four vertices is empty. For weighted points, the circumsphere is replaced by the *orthosphere*, which is the unique sphere orthogonal to all four spheres whose centers are the vertices of the tetrahedron. Its center is the Voronoi vertex shared by the four Voronoi regions, and its weight is the common weighted distance of that vertex from the four spheres. We summarize by generalizing the Circumcircle Claim of Section 1.1 to three dimensions and to the weighted case.

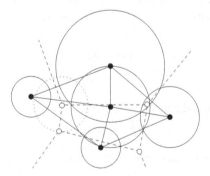

Figure 5.9. Dashed weighted Voronoi diagram and solid weighted Delaunay triangulation of five weighted points in the plane. Each Voronoi vertex is the center of a circle orthogonal to the generating circles of the regions that meet at that vertex. Only one such circle is shown.

Orthosphere Claim. A tetrahedron belongs to the weighted Delaunay tetra-hedrization if and only if the orthosphere of the four spheres is further than orthogonal from all other spheres in the set.

A sphere in S is *redundant* if its Voronoi region is empty. By definition, the center of a sphere is a vertex of the weighted Delaunay triangulation if and only if it is nonredundant. All extreme points are nonredundant, which implies that the underlying space is the convex hull of S, as in the unweighted case.

Local convexity

Recall the Delaunay Lemma of Section 1.2, which states that a triangulation of a finite set in $\mathbb{R}^2$ is the Delaunay triangulation if and only if every one of its edges is locally Delaunay. This result generalizes to three (and higher) dimensions and to the weighted case. For the purpose of this discussion, we define a *tetrahedrization* of S as a simplicial complex K whose underlying space is conv S and whose vertex set is a subset of S. A triangle abc in K is *locally convex* if

(i) it belongs to only one tetrahedron and therefore bounds the convex hull of S, or
(ii) it belongs to two tetrahedra, $abcd$ and $abce$, and $\hat{e}$ is further than orthogonal from the orthosphere of $abcd$.

If all triangles in K are locally convex, then after lifting we get the boundary complex of a convex polyhedron. This is consistent with the right side of the commuting diagram in Figure 5.7. However, to be sure this polyhedron is G, we also require that no lifted point lies vertically below the boundary.

Local Convexity Lemma. If Vert K contains all nonredundant weighted points and every triangle is locally convex, then K is the weighted Delaunay tetra-hedrization of S.

The proof is rather similar to that of the Delaunay Lemma in Section 1.2 and does not have to be repeated. Similarly, we can extend the Acyclicity Lemma of Section 1.1 to three (and higher) dimensions and to the weighted case. Details should be clear and are omitted.

Bibliographic notes

Weighted Voronoi diagrams are possibly as old as unweighted ones. Some of the earliest references appear in the context of quadratic forms, which arise in the study of the geometry of numbers [4]. These forms are naturally related to weighted as opposed to unweighted diagrams. Examples of such work are the papers by Dirichlet [2] and Voronoi [5]. Weighted Delaunay triangulations and their generalizations to three and higher dimensions seem less natural and have a shorter history. Nevertheless, they have already acquired at least three different names, namely regular triangulations [1] and coherent triangulations [3] besides the one used in this book.

[1] L. J. Billera and B. Sturmfels. Fiber polytopes. *Ann. Math.* **135** (1992), 527–549.
[2] P. G. L. Dirichlet. Über die Reduktion der positiven quadratischen Formen mit drei unbestimmten ganzen Zahlen. *J. Reine Angew. Math.* **40** (1850), 209–227.
[3] I. M. Gelfand, M. M. Kapranov, and A. V. Zelevinsky. *Discriminants, Resultants and Multidimensional Determinants.* Birkhäuser, Boston, 1994.
[4] P. M. Gruber and C. G. Lekkerkerker. *Geometry of Numbers.* Second edition, North-Holland, Amsterdam, 1987.
[5] G. Voronoi. Nouvelles applications des paramètres continus à la théorie des formes quadratiques. *J. Reine Angew. Math.* **133** (1907), 97–178, and **134** (1908), 198–287.

5.3 Flipping

The goal of this section is to generalize the idea of edge flipping to three and higher dimensions. We begin with two classic theorems in convex geometry. Helly's Theorem talks about the intersection structure of convex sets. It can be proved by using Radon's Theorem, which talks about partitions of finite point sets and is directly related to flips in d dimensions. We then define flips and discuss structural issues that arise in $\mathbb{R}^3$.

Radon's theorem

This is a result on $n \geq d + 2$ points in $\mathbb{R}^d$. The case of $n = 4$ points in $\mathbb{R}^2$ is related to edge flipping in the plane.

Radon's Theorem. Every collection S of $n \geq d + 2$ points in $\mathbb{R}^d$ has a partition $S = A \overset{.}{\cup} B$ with conv $A \cap$ conv $B \neq \emptyset$.

Proof. Since there are more than $d + 1$ points, they are affinely dependent. Hence there are coefficients λ_i, not all zero, with $\sum \lambda_i p_i = 0$ and $\sum \lambda_i = 0$.

Let I be the set of indices i with $\lambda_i > 0$, and let J contain all other indices. Note that $c = \sum_{i \in I} \lambda_i = -\sum_{j \in J} \lambda_j > 0$, and also

$$ x = \frac{1}{c} \cdot \sum_{i \in I} \lambda_i p_i = -\frac{1}{c} \cdot \sum_{j \in J} \lambda_j p_j. $$

Let A be the collection of points p_i with $i \in I$ and let B contain all other points. Point x is a convex combination of the points in A as well as of the points in B. Equivalently, $x \in \operatorname{conv} A \cap \operatorname{conv} B$. $\qquad \square$

A $(d + 1)$-dimensional simplex has $d + 2$ vertices and a face for every subset of the vertices. If we project its boundary complex onto $\mathbb{R}^d$ we get a simplex for every subset of at most $d + 1$ vertices. By Radon's Theorem, at least two of these simplices have an improper intersection. This intersection comes from projecting the two sides of the simplex boundary on top of each other.

Helly's theorem

This is a result on $n \geq d + 2$ convex sets in $\mathbb{R}^d$. For $d = 1$ it states that if every pair of a collection of $n \geq 2$ closed intervals has a nonempty intersection, then the entire collection has a nonempty common intersection. This is true because the premise implies that the rightmost left endpoint is to the left or equal to the leftmost right endpoint. The interval between these two endpoints belongs to every interval in the collection.

Helly's Theorem. If every $d + 1$ sets in a collection of $n \geq d + 2$ closed convex sets in $\mathbb{R}^d$ have a nonempty common intersection, then the entire collection has a nonempty intersection.

Proof. Assume inductively that the claim holds for $n - 1$ closed convex sets. For each C_i in the collection of n sets, let p_i be a point in the common intersection of the other $n - 1$ sets. Let S be the collection of points p_i. By Radon's Theorem, there is a partition $S = A \mathbin{\dot\cup} B$ and a point $x \in \operatorname{conv} A \cap \operatorname{conv} B$. By construction, $\operatorname{conv} A$ is contained in all sets C_j with $p_j \in B$, and symmetrically, $\operatorname{conv} B$ is contained in all sets C_i with $p_i \in A$. Hence, x is contained in every set of the collection. $\qquad \square$

Flipside of a simplex

Consider the case $d = 2$. The projection of a 3-simplex (tetrahedron) onto $\mathbb{R}^2$ is either a convex quadrangle or a triangle. In the former case the two diagonals

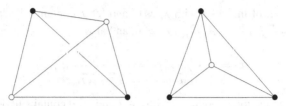

Figure 5.10. The two generic projections of a tetrahedron onto the plane.

cross, and in the latter case one vertex lies in the triangle spanned by the other three. Both cases are illustrated in Figure 5.10. The direction of projection defines an *upper* and a *lower* side of the tetrahedron boundary, and the two sides meet along the *silhouette*. Let $\alpha = \operatorname{conv} A$ and $\beta = \operatorname{conv} B$ be the two faces whose projections have an improper intersection. They lie on opposite sides, and we assume that α belongs to the upper and β to the lower side. The quadrangle case defines an edge flip, which replaces the projection of the upper by the projection of the lower side, or vice versa. We also call this a *two-to-two flip* because it replaces two old by two new triangles. The triangle case defines a new type of flip, which we refer to as a *one-to-three* or a *three-to-one flip*, depending on whether a new vertex is added or an old vertex is removed.

How do these considerations generalize to the case $d = 3$? As illustrated in Figure 5.11, the projection of a 4-simplex onto $\mathbb{R}^3$ is either a double pyramid or a tetrahedron. In the double pyramid case, α is an edge and β is a triangle. There are three tetrahedra that share α, and they form the upper side of the 4-simplex. The remaining two tetrahedra share β and form the lower side. The *three-to-two flip* replaces the projection of the upper side by the projection of the lower side, and the *two-to-three flip* does it the other way round. In the tetrahedron case, α is one vertex and β is the tetrahedron spanned by the other four vertices. The *one-to-four flip* adds α, effectively replacing β by four tetrahedra, and the *four-to-one flip* removes α.

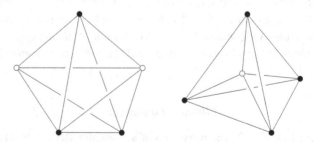

Figure 5.11. The two generic projections of a 4-simplex onto three-dimensional space.

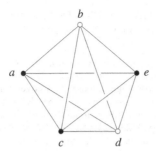

Figure 5.12. The edge *ae* does not pass through the triangle *bcd* but rather behind the edge *bd*.

Transformability

In using flips to construct a Delaunay tetrahedrization in $\mathbb{R}^3$, we encounter cases where we would like to flip but we cannot. This happens only for two-to-three flips. Let *abcd* and *bcde* share the triangle *bcd*. If the edge *ae* crosses *bcd*, we can replace *abcd*, *bcde* by *baec*, *caed*, *daeb*, which is a two-to-three flip. However, if the edge *ae* misses *bcd*, as illustrated in Figure 5.12 where *ae* passes behind *bd*, we cannot add *ae* because it might cross other triangles in the current tetrahedrization. In this case, the union of the two tetrahedra is nonconvex. Assume without loss of generality that *bd* is the nonconvex edge. There are two cases. If *bd* belongs to only three tetrahedra then the third one is *abde*, and we can replace *abdc*, *cbde*, *ebda* by *bace*, *aced*. This is a three-to-two flip. However, if *bd* belongs to four or more tetrahedra, then we are stuck and cannot remove the triangle *bcd*. This is the *nontransformable* case.

The reason for studying flips is of course the interest in an algorithm that constructs a weighted Delaunay tetrahedrization by flipping. The occurrence of nontransformable cases does not imply that all hope is lost. It might still be possible to flip elsewhere in a way that resolves nontransformable cases by changing their local neighborhood. However, this requires further analysis.

Bibliographic notes

Radon's Theorem is a byproduct of the effort by Johann Radon [4] to prove Helly's Theorem, communicated to him by Eduard Helly [1]. The two theorems are equivalent and form a cornerstone of modern convex geometry. Helly was missing as a prisoner of war in Russia, so Radon published his theorem and proof. After returning from Russia, Helly published his theorem and his own proof, which is inductive in the size of the collection *and* the dimension. Years later, Helly generalized his theorem to a topological setting where convexity is replaced by requirements of connectivity [2]. The concept of an edge flip was generalized to three and higher dimensions by Lawson [3], without, however, realizing the connection to Radon's Theorem.

[1] E. Helly. Über Mengen konvexer Körper mit gemeinschaftlichen Punkten. *Jahresber. Deutsch. Math.-Verein.* **32** (1923), 175–176.
[2] E. Helly. Über Systeme von abgeschlossenen Mengen mit gemeinschaftlichen Punkten. *Monatsh. Math. Physik* **37** (1930), 281–302.
[3] C. L. Lawson. Properties of n-dimensional triangulations. *Computer Aided Geometric Design* **3** (1986), 231–246.
[4] J. Radon. Mengen konvexer Körper, die einen gemeinschaftlichen Punkt enthalten. *Math. Ann.* **83** (1921), 113–115.

5.4 Incremental algorithm

This section generalizes the algorithm of Section 1.3 to three dimensions and to the weighted case. The algorithm is incremental and adds a point in a sequence of flips. We describe the algorithm, prove its correctness, and discuss its running time.

Algorithm

Let S be a finite set of weighted points in $\mathbb{R}^3$. We denote the points by $\hat{p}_1, \hat{p}_2, \ldots,$ $\hat{p}_n$ and assume they are in general position. To reduce the number of cases, we let $wxyz$ be a sufficiently large tetrahedron. In particular, we assume $wxyz$ contains all points of S in its interior. Define $S_i = \{w, x, y, z, \hat{p}_1, \hat{p}_2, \ldots, \hat{p}_i\}$ for $0 \leq i \leq n$, and let D_i be the weighted Delaunay tetrahedrization of S_i. The algorithm starts with D_0 and adds the weighted points in order. Adding $\hat{p}_i$ is done in a sequence of flips.

```
for i = 1 to n do
    find pqrs ∈ D_{i−1} that contains p_i;
    if p̂_i is non-redundant among p̂, q̂, r̂, ŝ then
        add p̂_i with a 1-to-4 flip
    endif;
    while ∃ triangle bcd not locally convex do
        flip bcd
    endwhile
endfor.
```

The algorithm maintains a tetrahedrization, which we denote as K. Sometimes, K is a weighted Delaunay tetrahedrization of a subset of the points, but often it is not. Consider flipping the triangle bcd in K. Let $abcd$ and $bcde$ be the two tetrahedra that share bcd. If their union is convex, then flipping bcd means a two-to-three flip that replaces bcd by edge ae together with triangles aeb, aec, aed.

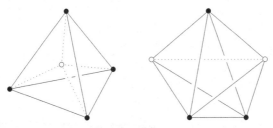

Figure 5.13. To the left, a one-to-four or a four-to-one flip depending on whether the hollow vertex is added or removed. To the right, a two-to-three or a three-to-two flip depending on whether the dotted edge is added or removed.

Otherwise, we consider the subcomplex induced by a, b, c, d, e. It consists of the simplices in K spanned by subsets of the five points. If the underlying space of the induced subcomplex is nonconvex, then bcd cannot be flipped. If the underlying space is convex, then it is either a double pyramid or a tetrahedron. In the former case, flipping means a three-to-two flip. In the latter case, flipping means a four-to-one flip, which effectively removes a vertex. The various types of flips are illustrated in Figure 5.13.

Stack of triangles

Flipping is done in a sequence controlled by a stack. At any moment, the stack contains all triangles in the link of p_i that are not locally convex. It may also contain other triangles in the link, but it contains each triangle at most once. Initially, the stack consists of the four triangles of $pqrs$. Flipping continues until the stack is empty.

```
while stack is non-empty do
    pop bcd from stack;
    if bcd ∈ K and bcd is not locally convex
                and bcd is transformable then
        apply a 2-to-3, 3-to-2, or 4-to-1 flip;
        push new link triangles on stack
    endif
endwhile.
```

Why can we restrict our attention to triangles in the link of p_i? Outside the link, K is equal to D_{i-1}; hence all triangles are locally convex. A triangle inside the link connects p_i with an edge cd in the link. Let xp_icd and p_icdy be the two tetrahedra sharing p_icd. If their union is convex, we can remove p_icd by a two-to-three flip. This creates a new tetrahedron $acde$ not incident to p_i, which

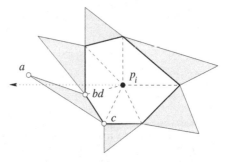

Figure 5.14. The bold edges belong to the link of p_i and the shaded triangles belong to L.

contradicts that D_{i-1} is a weighted Delaunay tetrahedrization. If their union is nonconvex, the triangles xcd and cdy in the link are also not locally convex.

Correctness

Let K be the tetrahedrization at some moment in time after adding $\hat{p}_i$ when it is not yet the weighted Delaunay tetrahedrization of S_i. It suffices to show that K has at least one link triangle that is not locally convex and transformable. To get a contradiction, we suppose all triangles that are not locally convex are nontransformable. Let L be the set of tetrahedra in $K - \operatorname{St} p_i$ that have at least one triangle in the link. These tetrahedra form a spiky sphere around p_i, not unlike the spiky circle in Figure 5.14. Let $L' \subseteq L$ contain all tetrahedra whose triangles in the link are not locally convex. By assumption, $L' \neq \emptyset$. For each tetrahedron in L, consider the orthosphere $\hat{z}$ and the weighted distance $\pi_{\hat{p}_i, \hat{z}}$. Let $abcd \in L$ be the tetrahedron whose orthosphere minimizes that function. We have $abcd \in L'$, or equivalently $\pi_{\hat{p}_i, \hat{z}} < 0$, for else the triangle bcd in the link would be locally convex, and so would every other link triangle.

We argue that bcd is transformable. To get a contradiction, assume it is not. Let bd be a nonconvex edge of the union of $abcd$ and $bcdp_i$, and let $abdx$ be the tetrahedron on the other side of abd. If bd is the only nonconvex edge then $x \neq p_i$, for else bcd would be transformable. Otherwise, there is another nonconvex edge, say bc. Let $abcy$ be the tetrahedron on the other side of abc. If $x = y = p_i$ we again have a contradiction because this would imply that bcd is transformable. We may therefore assume that $x \neq p_i$. Equivalently, abd is not in the link of $\hat{p}_i$. Consider a half-line that starts at p_i and passes through an interior point of abd. After crossing the link, the half-line goes through a tetrahedron of L before it encounters $abcd$. This is illustrated in Figure 5.14. Outside the link, we have a genuine weighted Delaunay tetrahedrization, namely a portion of D_{i-1}. For tetrahedra in D_{i-1}, the weighted distance of $\hat{p}_i$ from their orthospheres increases along the half-line, which contradicts the minimality assumption in the choice of *abcd*. This finally proves that flipping continues until D_i is reached.

Number of flips

To upper-bound the number of flips in the worst case, we interpret the algorithm as gluing 4-simplices to a three-dimensional surface consisting of tetrahedra in $\mathbb{R}^4$. Each flip corresponds to a 4-simplex. It either removes or introduces one or four edges. Once an edge is removed it cannot be introduced again. This implies that the total number of flips is less than $2\binom{n}{2} < n^2$. We thus have an algorithm that constructs the Delaunay tetrahedrization of n points in $\mathbb{R}^3$ in $O(n^2)$ time. The size of the final Delaunay tetrahedrization is therefore at most some constant times n^2.

There are sets of n points in $\mathbb{R}^3$ with at least some constant times n^2 Delaunay tetrahedra. Take, for example, two skew lines and place $n/2$ unweighted points on each line, as shown in Figure 5.15. Consider two contiguous points on one line together with two contiguous points on the other line. The sphere passing through the four points is empty, which implies that the four points span a Delaunay tetrahedron. The total number of such tetrahedra is roughly $n^2/4$. However, for point sets that seem to occur in practice, the number of Delaunay tetrahedra is typically less than some constant times n. Examples of such sets are dense packing of spheres common in molecular modeling, and well-spaced sets as produced by three-dimensional mesh generation software.

Expected running time

It is a good idea to first compute a random permutation of the points so that the construction proceeds in a random order. However, because the size of the

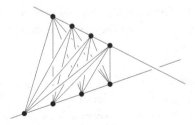

Figure 5.15. A tetrahedral mesh whose edge skeleton contains a complete bipartite graph.

tetrahedrization can vary between linearly and quadratically many simplices, the analysis is more involved than in two dimensions. We cannot even claim that the expected running time is at most $\log_2 n$ times the size of the final tetrahedrization. Indeed, this is false because there exist point sets with linear size Delaunay tetrahedrizations that reach quadratic intermediate size with positive constant probability. Nevertheless, such a claim holds if we further relativize the statement by drawing points from a fixed distribution. Suppose the expected size of the Delaunay tetrahedrization of k points chosen randomly from the distribution is $O(f(k))$. If $f(k) = \Omega(k^{1+\varepsilon})$, for some constant $\varepsilon > 0$, then the expected running time is $O(f(n))$; otherwise it is $O(f(n) \log n)$. The argument is similar to the one presented in Section 1.3, and details are omitted.

Bibliographic notes

Algorithms that construct a Delaunay tetrahedrization in $\mathbb{R}^3$ through flips were first considered by Barry Joe. In [2] he gives an example in which the non-transformable cases form a deadlock situation and flipping does not lead to the Delaunay tetrahedrization. In [3] he shows that flipping succeeds if the points are added one at a time. The proof of Joe's result in this section is taken from [1], where the same is shown for weighted Delaunay tetrahedrization in $\mathbb{R}^d$.

[1] H. Edelsbrunner and N. R. Shah. Incremental topological flipping works for regular triangulations. *Algorithmica* **15** (1996), 223–241.
[2] B. Joe. Three-dimensional triangulations from local transformations. *SIAM J. Sci. Statist. Comput.* **10** (1989), 718–741.
[3] B. Joe. Construction of three-dimensional Delaunay triangulations from local transformations. *Comput. Aided Geom. Design* **8** (1991), 123–142.

Exercise collection

The credit assignment reflects a subjective assessment of difficulty. A typical question can be answered by using knowledge of the material combined with some thought and analysis.

1. **Inscribed polytopes** (three credits). A 3-polytope *inscribed* in the two-dimensional sphere has all its vertices on the sphere. Prove that the cube with one corner cut off cannot be inscribed. We permit geometric distortions of the cube, but edges and facets must be straight and the combinatorial structure must be the same as that in Figure 5.16.

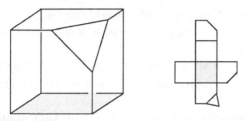

Figure 5.16. Cube with one corner cut off and its net.

2. **Helly for rectangles** (two credits). Define a *rectangle* in $\mathbb{R}^2$ as the set of points $x = (x_1, x_2)$, with $\ell_i \leq x_i \leq r_i$ for some real numbers ℓ_i, r_i and $i = 1, 2$. Let R be a finite collection of rectangles.
   (i) Prove if every pair of rectangles has a nonempty intersection then $\bigcap R \neq \emptyset$.
   (ii) Generalize rectangles and the claim in (i) from two to three and higher dimensions.

3. **Jung's theorem** (two credits). A theorem by Jung states that if every three of a finite set of points in the plane are contained in a unit disk, then the entire set is contained in a unit disk.
   (i) Use the two-dimensional version of Helly's Theorem to prove Jung's Theorem.
   (ii) What is the generalization of Jung's Theorem to $d \geq 3$ dimensions?

4. **Degenerate Delaunay complex** (two credits). The *face-centered cube (FCC) lattice* consists of all integer points (i, j, k) with even sum $i + j + k$.
   (i) The Delaunay complex of the FCC lattice is degenerate because there are empty spheres that pass through six points. Which six points?
   (ii) The Delaunay complex of the FCC lattice has only two types of three-dimensional cells. What are they?

5. **Non-Delaunay complex** (one credit). Exhibit a three-dimensional simplicial complex K with vertex set $S \subseteq \mathbb{R}^3$ whose edges all belong to the Delaunay tetrahedrization D of S and whose underlying space is the convex hull of S, but still $K \neq D$.

6. **Induced subcomplex** (three credits). Let S be a finite set of points in $\mathbb{R}^3$ and D the Delaunay tetrahedrization of S. The subcomplex $K \subseteq D$ *induced* by a subset $T \subseteq S$ consists of all simplices in D whose vertices all belong to T.
   (i) Show that K is also a subcomplex of the Delaunay tetrahedrization of T.
   (ii) Use (i) to show that if there is a Delaunay tetrahedrization that has an edge crossing a triangle, then there is such a Delaunay tetrahedrization for only five points.

7. **Moment curve** (three credits). The *moment curve* in $\mathbb{R}^3$ consists of all points (t, t^2, t^3) with $t \in \mathbb{R}$. Let $p_1, p_2, \ldots p_n$ be a sequence of points along the moment curve.

  (i) Show that for all $1 < i < j < n$ the sphere passing through points $p_{i-1}, p_i, p_j, p_{j+1}$ is empty. In other words, all other points p_ℓ lie outside that sphere.

  (ii) Count the tetrahedra, triangles, edges of the Delaunay tetrahedrization of $p_1, p_2, \ldots, p_n$.

8. **Local convexity** (two credits). Let ab be an edge in a tetrahedrization K of $S \subseteq \mathbb{R}^3$. Prove that if ab belongs to a triangle that is not locally convex, then it belongs to at least three such triangles.

6

Tetrahedron meshes

This chapter studies the problem of constructing meshes of tetrahedra in $\mathbb{R}^3$. Such meshes are three-dimensional simplicial complexes, the same as what we called tetrahedrizations in Chapter 5. The new aspects are the attention to boundary conditions and the focus on the shape of the tetrahedra. The primary purpose of meshes is to provide a discrete representation of continuous space. The tetrahedra themselves and their arrangement within the mesh are not as important as how well they represent space. Unfortunately, there is no universal measure that distinguishes good from bad space representations. As a general guideline, we avoid very small and very large angles because of their usually negative influence on the performance of numerical methods based on meshes. Section 6.1 studies the problem of tetrahedrizing possibly nonconvex polyhedra. Section 6.2 measures tetrahedral shape and introduces the ratio property for Delaunay tetrahedrizations. Section 6.3 extends the Delaunay refinement algorithm from two to three dimensions. Section 6.4 studies a particularly annoying type of tetrahedron and ways to remove it from Delaunay meshes.

6.1 Meshing polyhedra

In this book, meshing a spatial domain means decomposing a polyhedron into tetrahedra that form a simplicial complex. This section introduces polyhedra and studies the problem of how many tetrahedra are needed to mesh them.

Polyhedra and faces

A *polyhedron* is the union of convex polyhedra, $P = \bigcup_{i \in I} \bigcap H_i$, where I is a finite index set and each H_i is a finite set of closed half-spaces. For example, the polyhedron in Figure 6.1 can be specified as the union of four convex polyhedra.

111

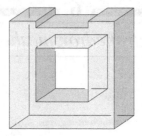

Figure 6.1. A nonconvex polyhedron.

As we can see, faces are not necessarily simply connected. We use a definition that permits faces even to be disconnected.

Let b be the open ball with unit radius centered at the origin of $\mathbb{R}^3$. For a point x we consider a sufficiently small neighborhood, $N_\varepsilon(x) = (x + \varepsilon \cdot b) \cap P$. The *face figure* of x is the enlarged version of this neighborhood within the polyhedron, $x + \bigcup_{\lambda > 0} \lambda \cdot (N_\varepsilon(x) - x)$. A *face* of P is the closure of a maximal collection of points with identical face figures. To distinguish the faces of P from the edges and triangles of the Delaunay tetrahedrization to be constructed, we call 1- and 2-faces of P *segments* and *facets*. Observe that the polyhedron in Figure 6.1 has 24 vertices, 30 segments, 11 facets, and two 3-faces, namely the inside with face figure $\mathbb{R}^3$ and the outside with an empty face figure. Six of the segments and three of the facets are non-connected. Two of the facets are connected but not simply connected, namely the front and the back facets.

Tetrahedrizations

A *tetrahedrization* of P is a simplicial complex K whose underlying space is P, $|K| = P$. Since simplicial complexes are finite by definition, only bounded polyhedra have tetrahedrizations. A tetrahedrization of P triangulates every facet and every segment by a subcomplex each. Every vertex of P is necessarily also a vertex of K.

We will see shortly that every bounded polyhedron has a tetrahedrization. Interestingly, there are polyhedra whose tetrahedrizations have necessarily more vertices than the polyhedra. The smallest such example is the Schönhardt polyhedron shown in Figure 6.2. It can be obtained from a triangular prism by a slight rotation of one triangular facet relative to the other. The six vertices of the polyhedron span $\binom{6}{4} = 15$ tetrahedra, which we classify into three types exemplified by *abcA*, *abAB*, *bcCA*. All three tetrahedra share *bA* as an edge. But this edge lies outside the Schönhardt polyhedron, which implies that none of the 15 tetrahedra is contained in the polyhedron. The Schönhardt polyhedron

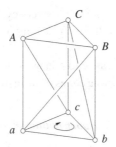

Figure 6.2. The Schönhardt polyhedron. The edges aB, bC, cA are non-convex.

can therefore not be tetrahedrized by using tetrahedra spanned by its vertices. There are, of course, other tetrahedrizations. The simplest uses a vertex z in the center and cones from z to the six vertices, 12 edges, and eight triangles in the boundary.

Fencing off

We give a constructive proof that every polyhedron P has a tetrahedrization. For simplicity we assume that P is everywhere three-dimensional. Equivalently, P is the closure of its interior, $P = \text{cl int } P$. It is convenient to place P in space such that no facet lies in a vertical plane and no segment is contained in a vertical line. Call two points $x, y \in P$ *vertically visible* if x, y lie on a common vertical line and the edge xy is contained in P. The *fence* of a segment consists of all points $x \in P$ vertically visible from some point y of the segment. The tetrahedrization is constructed in three steps, the first of which is illustrated in Figure 6.3.

Step 1. Erect the fence of each segment. The fences decompose P into vertical cylinders, each bounded by a top and a bottom facet and a circle of fence pieces called *walls*.

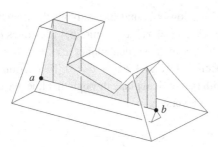

Figure 6.3. The fence of the segment ab consists of five walls, each a triangle or a quadrangle.

`Step 2.` Triangulate the bottom facet of every cylinder and erect fences from the new segments, effectively decomposing P into triangular cylinders.

`Step 3.` Decompose each wall into triangles and finally tetrahedrize each cylinder by constructing cones from an interior point to the boundary.

Upper bound

We analyze the tetrahedrization obtained by erecting fences and prove that the final number of tetrahedra is at most some constant times the square of the number of segments.

Upper Bound Claim. The three steps tetrahedrize a bounded polyhedron with m segments by using fewer than $28m^2$ tetrahedra.

Proof. Fences erected in `Step 1` may meet in vertical edges. Each intersection corresponds to a crossing between vertical projections of segments. The total number of crossings is at most $\binom{m}{2}$. Each segment creates a fence, and each crossing involving this segment may cut one wall of the fence into two. The total number of walls is therefore no more than $m + 2\binom{m}{2} = m^2$. A cylinder bounded by k walls is decomposed into $k - 2$ triangular cylinders separated from each other by $k - 3$ new walls. `Step 2` thus increases the total number of walls to less than $3m^2$. The total number of cylinders at this stage is less than $2m^2$. Each wall is a triangle or a quadrangle, and it may be divided into two by the piece of the segment that defines it. `Step 3` therefore triangulates each wall by using four or fewer triangles, and it tetrahedrizes each cylinder by using 14 or fewer tetrahedra. The final tetrahedrization thus contains fewer than $28m^2$ tetrahedra. $\square$

Saddle surface

We prepare a matching lower bound by studying the *hyperbolic paraboloid* specified by the equation $x_3 = x_1 \cdot x_2$. Figure 6.4 illustrates the paraboloid by showing its intersection with the vertical planes $\pm x_1 \pm x_2 = 1$. A general line in the $x_1 x_2$-plane is specified by $ax_1 + bx_2 + c = 0$. To determine the intersection of the paraboloid with the vertical plane through that line, we can either express x_1 in terms of x_2 or vice versa:

$$x_3 = -\frac{b}{a}x_2^2 - \frac{c}{a}x_2,$$

$$x_3 = -\frac{a}{b}x_1^2 - \frac{c}{b}x_1.$$

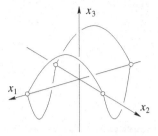

Figure 6.4. Hyperbolic paraboloid indicated through its intersection with vertical walls.

For $a \cdot b \neq 0$ we get a parabola. For $a = 0$ we get a line for every value of c/b, and we sample this family at integer values. Similarly, we sample the 1-parameter family of lines we get for $b = 0$ at integer values of c/a. Figure 6.5 shows a small portion of the two families in top view. If two points x and y lie on the paraboloid, then the segment between them lies on the surface if and only if the vertical projections of x, y onto the $x_1 x_2$-plane line on a common horizontal or vertical line. If the line has positive slope then the segment lies above the surface, and if the line has negative slope then it lies below the surface.

Lower bound construction

We build a polyhedron Q out of a cube by cutting deep wedges, each close to a line of the two ruling families. The construction is illustrated in Figure 6.6. Assuming we have n cuts from the top and n from the bottom, we have $m = 14n + 8$ segments forming the polyhedron.

Lower Bound Claim. Every tetrahedrization of Q consists of at least $(n + 1)^2$ tetrahedra.

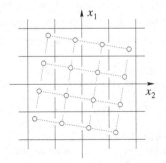

Figure 6.5. View from below of hyperbolic paraboloid. We see samples of the two ruling families of lines and dotted edges connecting points sampled on the surface.

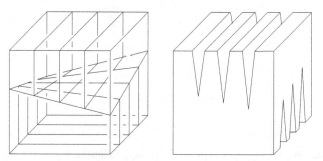

Figure 6.6. Polyhedron Q with two families of cuts almost meeting along the saddle surface.

Proof. Consider the checkerboard produced by the $2n + 4$ lines on the saddle surface that mark the ends of the $2n$ cuts and the intersection with the boundary of the cube. Choose a point in each square of the checkerboard producing the slightly tilted square grid pattern of Figure 6.5. The edges connecting any two points intersect at least one of the wedges, provided the sharp ends of the wedges reach sufficiently close to the saddle surface. It follows that in any tetrahedrization of Q, the $(n + 1)^2$ points lie inside pairwise different tetrahedra. □

Bibliographic notes

The definition of a polyhedron as the union of intersections of closed half-spaces is taken from Hadwiger [4]. The definition of a face is taken from Edelsbrunner [2] and should be contrasted with that suggestion in [3]. The Schönhardt polyhedron was named after E. Schönhardt, who described the polyhedron in 1928 [7]. The same construction was mentioned 17 years earlier in a paper by Lennes [5]. Ruppert and Seidel build on this construction, and show that deciding whether or not a polyhedron can be tetrahedrized without adding new vertices is NP-complete [6]. The quadratic upper and lower bounds for tetrahedrizing polyhedra are taken from a paper by Bernard Chazelle [1].

[1] B. Chazelle. Convex partitions of polyhedra: a lower bound and worst case algorithm. *SIAM J. Comput.* **13** (1984), 488–507.
[2] H. Edelsbrunner. Algebraic decomposition of non-convex polyhedra. In "Proc. 36th Ann. IEEE Sympos. Found. Comput. Sci.," 1995, 248–257.
[3] B. Grünbaum and G. C. Shephard. A new look at Euler's theorem for polyhedra. *Amer. Math. Monthly* **101** (1994), 109–128.
[4] H. Hadwiger. *Vorlesungen über Inhalt, Oberfläche und Isoperimetrie.* Springer, Berlin, 1957.
[5] N. J. Lennes. Theorems on the simple finite polygon and polyhedron. *Amer. J. Math.* **33** (1911), 37–62.

[6] J. Ruppert and R. Seidel. On the difficulty of triangulating three-dimensional non-convex polyhedra. *Discrete Comput. Geom.* **7** (1992), 227–254.

[7] E. Schönhardt. Über die Zerlegung von Dreieckspolyedern in Tetraeder. *Math. Ann.* **98** (1928), 309–312.

6.2 Tetrahedral shape

This section looks at the various shapes tetrahedra can assume. For the time being, good shape quality is defined as having a small circumradius over the shortest edge length ratio. We will see later that meshes of tetrahedra with small ratios also have nice combinatorial properties, such as constant size vertex stars.

Classifying tetrahedra

The classification of tetrahedra into shape types is a fuzzy undertaking. We normalize by scaling tetrahedra to unit diameter. A normalized tetrahedron has small volume either because its vertices are close to a line, or, if that is not the case, its vertices are close to a plane. In the first case, the tetrahedron is *skinny*, and we distinguish five types depending on how its vertices cluster along the line. Up to symmetry, the possibilities are 1-1-1-1, 1-1-2, 1-2-1, 1-3, 2-2, as shown from left to right in Figure 6.7. A *flat* tetrahedron has small volume but is not skinny. We have four types depending on whether two vertices are close to each other, three vertices lie close to a line, the orthogonal projection of the tetrahedron onto the close plane is a triangle, or the projection is a quadrangle. All four types are shown from left to right in Figure 6.8.

Circumradius over the shortest edge length

A tetrahedron $abcd$ has a unique circumsphere. Let $R = R(abcd)$ be that radius and $L = L(abcd)$ the length of the shortest edge. We measure the quality of the

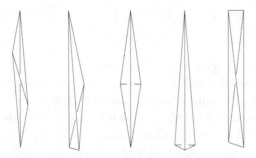

Figure 6.7. Five fuzzy types of skinny tetrahedra.

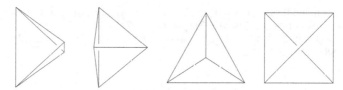

Figure 6.8. Four fuzzy types of flat tetrahedra, the fourth is the sliver.

tetrahedron shape by taking the ratio, that is,

$$\varrho = \varrho(abcd) = R/L.$$

We also define ϱ for triangles, taking the radius of the circumcircle over the length of the shortest edge. Observe that the ratio of a tetrahedron is always larger than or equal to the ratio of each of its triangles.

A triangle abc minimizes the ratio if and only if it is equilateral, in which case the circumcenter is also the barycenter,

$$y = 1/3 \cdot (a + b + c) = 2/3 \cdot x + 1/3 \cdot c,$$

where $x = 1/2 \cdot (a + b)$. Normalization implies that the three edges have length 1. The ratio is therefore equal to the circumradius, which is

$$\|c - y\| = 2/3 \cdot \|c - x\| = 2/3 \cdot \sqrt{1 - 1/4}$$
$$= \sqrt{3}/3 = 0.577\ldots.$$

A tetrahedron $abcd$ minimizes the ratio if and only if it is regular, in which case the circumcenter is again the barycenter:

$$z = 1/4 \cdot (a + b + c + d) = 3/4 \cdot y + 1/4 \cdot d.$$

Normalization implies that the six edges have length 1. The ratio is therefore equal to the circumradius, which is

$$\|d - z\| = 3/4 \cdot \|d - y\| = 3/4 \cdot \sqrt{1 - 3/9}$$
$$= \sqrt{6}/4 = 0.612\ldots.$$

Both calculations are illustrated in Figure 6.9.

A *skinny* triangle has small area. It either has a short edge or a large circumradius. In either case, its ratio is large. A skinny tetrahedron has skinny triangles; hence its ratio is large. A flat triangle that is not a sliver has either a short edge or a large circumradius and thus a large ratio. The only remaining small volume tetrahedron is the sliver, and it can have ϱ as small as $\sqrt{2}/2 = 0.707\ldots$, or even a tiny amount smaller.

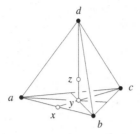

Figure 6.9. A regular tetrahedron and the barycenters of an edge, a triangle, the tetrahedron.

Ratio property

A mesh of tetrahedra has the *ratio property for* ϱ_0 if $\varrho \leq \varrho_0$ for all tetrahedra. We assume that every triangle in the mesh is the face of a tetrahedron in the mesh. It follows that $\varrho \leq \varrho_0$ also for every triangle. We prove two elementary facts about edge lengths in a mesh K that has the ratio property for a constant ϱ_0.

Claim A. If abc is a triangle in K then

$$1/2\varrho_0 \cdot \|a - b\| \leq \|a - c\| \leq 2\varrho_0 \cdot \|a - b\|.$$

Proof. The length of an edge is at most twice the circumradius, $\|a - b\| \leq 2Y$. By assumption, $\|a - b\| \geq Y/\varrho_0$. The same inequalities hold for $\|a - c\|$, which implies the claim. □

Next we show that, if K has the ratio property and it is a Delaunay tetrahedrization, then edges that share a common endpoint and form a small angle cannot have very different lengths. For this to hold, it is not necessary that the two edges belong to a common triangle. Define

$$\eta_0 = \arctan 2\left(\varrho_0 - \sqrt{\varrho_0^2 - 1/4}\right).$$

Since ϱ_0 is a constant, so is η_0.

Claim B. If the angle between ab and ap is less than η_0 then

$$\|a - b\|/2 < \|a - p\| < 2 \cdot \|a - b\|.$$

Proof. Consider the circumsphere of a tetrahedron that contains ab as an edge, and let $\hat{y} = (y, Y^2)$ be the circle in which the plane passing through a, b, p intersects the sphere. We use Figure 6.10 as an illustration throughout the proof.

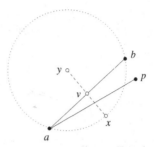

Figure 6.10. Section through a circumsphere of a Delaunay tetrahedron with edge ab.

Let v be the midpoint of ab, and let x be the point on the circle such that y, v, x lie in this sequence on a common line. We have $Y \leq \varrho_0 \cdot \|a - b\|$ by assumption. The distance between x and v is

$$\|x - v\| = Y - \sqrt{Y^2 - \|a - b\|^2/4}$$
$$\geq \left(\varrho_0 - \sqrt{\varrho_0^2 - 1/4}\right) \cdot \|a - b\|,$$

because the difference between Y and $\sqrt{Y^2 - C}$ decreases with increasing Y. The angle between ab and ax is

$$\angle bax = \arctan(2\|x - v\|/\|a - b\|)$$
$$\geq \arctan 2\left(\varrho_0 - \sqrt{\varrho_0^2 - 1/4}\right)$$
$$= \eta_0.$$

The claimed lower bound follows because the circle forces ap to be at least as long as ax, which is longer than half of ab. The claimed upper bound on the length of ap follows by a symmetric argument that reverses the roles of b and p.   □

Length variation

We use Claims A and B to show that the length variation of edges with a common endpoint a in K is bounded by some constant. As before, we assume K has the ratio property and is a Delaunay tetrahedrization. Define $m_0 = 2/(1 - \cos(\eta_0/4))$ and $\nu_0 = 2^{2m_0-1} \cdot \varrho_0^{m_0-1}$. Since ϱ_0 and η_0 are constants, so are m_0 and ν_0.

Length Variation Lemma. If ab, ap are edges in K then

$$\|a - b\|/\nu_0 < \|a - p\| < \nu_0 \cdot \|a - b\|.$$

Proof. Let Σ be the sphere of directions around a. We form a maximal packing of circular caps, each with angle $\eta_0/4$. This means if y is the center and x a boundary point of a cap then $4\angle xay = \eta_0$. The area of each cap is $(1 - \cos(\eta_0/4))/2$ times the area of Σ, which implies that there are at most m_0 caps.

By increasing the caps to radius $\eta_0/2$ we change the maximal packing into a covering of Σ. For each edge ab in the star of a, let $b' \in \Sigma$ be the radial projection of b. Similarly, for each triangle abc consider the arc on Σ that is the radial projection of bc. The points and arcs form a planar graph. Let ab be the longest and ap the shortest edge in the star of a. We walk in the graph from b' to p'. This path leads from cap to cap, and we record the sequence ignoring detours that return to previously visited caps. The sequence consists of at most m_0 caps. Let us track the edge length during the walk. As long as we stay within a cap, Claim B implies the length decreases by less than a factor $1/2$. If we step from one cap to the next, Claim A implies the length decreases by at most a factor $1/2\varrho_0$. Hence $\|a - p\| > \|a - b\|/v_0$. The upper bound follows by a symmetric argument that exchanges b and p. $\square$

Constant degree

A straightforward volume argument together with the Length Variation Lemma implies that each vertex in K belongs to at most some constant number of edges. Define $\delta_0 = (2v_0^2 + 1)^3$, which is a constant.

Degree Lemma. Every vertex a in K belongs to at most δ_0 edges.

Proof. Let ab be the longest and ap the shortest edge in the star of a. Assume without loss of generality that $\|a - p\| = 1$. Let c be a neighbor of a and let d be a neighbor of c. We have $\|a - c\| \geq 1$ by assumption and $\|c - d\| \geq 1/v_0$ by the Length Variation Lemma. For each neighbor c of a let Γ_c be the open ball with center c and radius $1/2v_0$. The balls are pairwise disjoint and fit inside the ball Γ with center a and radius $\|a - b\| + 1/2v_0$. The volume of Γ is

$$
\begin{aligned}
\text{vol}\,\Gamma &= \frac{4\pi}{3}\left(\|a - b\| + \frac{1}{2v_0}\right)^3 \\
&\leq \frac{4\pi}{3}\left(\frac{2v_0^2 + 1}{2v_0}\right)^3 \\
&= \left(2v_0^2 + 1\right)^3 \cdot \text{vol}\,\Gamma_c.
\end{aligned}
$$

In words, at most $\delta_0 = (2v_0^2 + 1)^3$ neighbor balls fit into Γ. This implies that δ_0 is an upper bound on the number of neighbors of a. $\square$

The constant δ_0 in the Degree Lemma is miserably large. The main reason is that the constant ν_0 in the Length Variation Lemma is miserably large. It would be nice to find a possibly more direct proof of that lemma and bring the constant down to a reasonable size.

Bibliographic notes

The idea of measuring the quality of a tetrahedron by its circumradius over the shortest edge length ratio is from Miller and coauthors [1]. The proofs of the Length Variation and Degree Lemmas are taken from the same source. Further results on meshes of tetrahedra that have the ratio property can be found in the doctoral thesis by Talmor [2].

[1] G. L. Miller, D. Talmor, S.-H. Teng, and N. Walkington. A Delaunay based numerical method for three dimensions: generation, formulation, and partition. In "Proc. 27th Ann. ACM Sympos. Theory Comput.," 1995, 683–692.
[2] D. Talmor. Well-spaced points for numerical methods. Report CMU-CS-97-164, Dept. Comput. Sci., Carnegie-Mellon Univ., Pittsburgh, Pennsylvania. 1997.

6.3 Delaunay refinement

This section generalizes the Delaunay refinement algorithm of Section 2.2 from two to three dimensions. The additional dimension complicates matters. In particular, special care must be taken to avoid infinite loops bouncing back and forth between refining segments and facets of the input polyhedron.

Refinement algorithm

For technical reasons, we restrict ourselves to bounded polyhedra P without interior angles smaller than $\pi/2$. The condition applies to angles between two segments, between a segment and a facet, and between two facets. The polyhedron in Figure 6.1 satisfies the condition, but the polyhedron in Figure 6.2 does not. The goal is to construct a Delaunay tetrahedrization D with a subcomplex $K \subseteq D$ that subdivides P and has the ratio property for a constant ϱ_0. The first step of the algorithm computes D as the Delaunay tetrahedrization of the set of vertices of P. Unless we are lucky, there will be segments that are not covered by edges of D, and there will be facets that are not covered by triangles of D. To recover these segments and facets, we add new points and update the Delaunay tetrahedrization using the incremental algorithm of Section 5.4. The points are added using the three rules given below.

We need some definitions. A segment of P is decomposed into *subsegments* by vertices of the Delaunay tetrahedrization that lie on the segment, and a facet is decomposed into (triangular) *subfacets* by the Delaunay triangulation of the vertices on the facet and its boundary. A vertex *encroaches upon* a subsegment if it is enclosed by the diameter sphere of that subsegment, and it *encroaches upon* a subfacet if it is enclosed by the equator sphere of that subfacet. Both spheres are the smallest that pass through all vertices of the subsegment and the subfacet.

Rule 1. If a subsegment is encroached upon, we split it by adding the midpoint as a new vertex to the Delaunay tetrahedrization. The new subsegments may or may not be encroached upon, and splitting continues until none of the subsegments is encroached upon.

Rule 2. If a subfacet is encroached upon, we split it by adding the circumcenter x as a new vertex to the Delaunay tetrahedrization. However, if x encroaches upon one or more subsegments then we do not add x and instead split the subsegments.

Rule 3. If a tetrahedron inside P has a circumradius over the shortest edge length ratio $R/L > \varrho_0$, then we split the tetrahedron by adding the circumcenter x as a new vertex to the Delaunay tetrahedrization. However, if x encroaches upon any subsegments or subfacets, we do not add x and instead split the subsegments and subfacets.

Rule 1 takes priority over **Rule 2**, and **Rule 2** takes priority over **Rule 3**. At the time we add a point on a facet, the prioritization guarantees that the boundary segments of the facet are subdivided by edges of the Delaunay tetrahedrization. Similarly, at the time we add a point in the interior of P, the boundary of P is subdivided by triangles in the Delaunay tetrahedrization. A point considered for addition to the Delaunay tetrahedrization has a *type*, which is the number of the rule that considers it or equivalently the dimension of the simplex it splits. Points of type 1 split subsegments and are always added once they are considered. Points of type 2 and 3 may be added or rejected.

Local density

Just as in two dimensions, the *local feature size* is crucial to understanding the Delaunay refinement algorithm. It is the function $f : \mathbb{R}^3 \to \mathbb{R}$ with $f(x)$ the radius of the smallest closed ball with center x that intersects at least two disjoint faces of P. Note that f is bounded away from zero by some positive constant. It is easy to show that f satisfies the Lipschitz condition

$$f(x) \le f(y) + \|x - y\|.$$

This implies that f is continuous over $\mathbb{R}^3$, but more than that, the condition says that f varies only slowly with x.

The local feature size is related to the *insertion radius* r_x of a point x, which is the length of the shortest Delaunay edge with endpoint x immediately after adding x. If x is a vertex of P then r_x is the distance to the nearest other vertex of P. If x has type 1 or 2 then r_x is the distance to the nearest encroaching vertex. If that encroaching vertex does not exist because it was rejected, then r_x is either half the length of the subsegment if x has type 1, or it is the circumradius of the subfacet if x has type 2. Finally, r_x is the circumradius of the tetrahedron it splits, if x has type 3. We also define the insertion radius for a point that is considered for addition but rejected, because it encroaches upon subsegments or subfacets. This is done by hypothetically adding the point and taking the length of the shortest edge in the hypothetical star.

Radii and parents

Points are added in a sequence, and for each new point there are predecessors that we can make responsible for the addition. If x has type 1 or 2 then we define the responsible *parent* $p = p_x$ as the encroaching point that triggers the event. The point p may be a Delaunay vertex or a rejected circumcenter. If there are several encroaching points then p is the one closest to x. If x has type 3 then p is the most recently added endpoint of the shortest edge of the tetrahedron x splits.

Radius Claim. Let x be a vertex of D and p its parent, if it exists. Then $r_x \geq f(x)$ or $r_x \geq c \cdot r_p$, where $c = 1/\sqrt{2}$ if x has type 1 or 2 and $c = \varrho_0$ if x has type 3.

Proof. If x is a vertex of P then $f(x)$ is less than or equal to the distance to the nearest other vertex. This distance is $r_x \geq f(x)$. For the rest of the proof assume x is not a vertex of P. It therefore has a parent $p = p_x$. First consider the case where p is a vertex of P. If x has type 1 or 2, it lies in a segment or facet of P, and p is not contained in that segment or facet. Hence $r_x = \|x - p\| \geq f(x)$. If x has type 3 then the tetrahedron split by x has at least two vertices in P. Hence $r_x = \|x - p\| \geq f(x)$ as before. Second consider the case where p is not a vertex of P. If x has type 1 or 2 then p was rejected for triggering the insertion of x. Since p encroaches upon the subsegment or subfacet split by x, its distance to the closest vertex of that subsegment or subfacet is at most $\sqrt{2}$ times the distance of x from that same vertex. Hence $r_x \geq r_p/\sqrt{2}$. Finally, if x has type 3 then $r_p \leq L$, where L is the length of the shortest edge of the

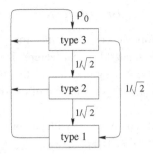

Figure 6.11. The directed arcs indicate possible parent–child relations, and their labels give the worst-case factors relating insertion radii.

tetrahedron split by x. The algorithm splits that tetrahedron only if $R > L\varrho_0$. Hence $r_x = R > L\varrho_0 \geq \varrho_0 r_p$.   □

Termination

The Radius Claim limits how quickly the insertion radius can decrease. We aim at choosing the only independent constant, which is ϱ_0, such that the insertion radii are bounded from below by a positive constant. Once this is achieved, we can prove termination of the algorithm by using a standard packing argument. Figure 6.11 illustrates the possible parent–child relations between the three types of points added by the algorithm. We follow an arc of the digraph whenever the insertion radius of a point x is less than $f(x)$. The arc is labeled by the smallest possible factor relating the insertion radius of x to that of its parent. Note that there is no arc from type 1 to type 2 and there are no loops from type 1 back to type 1 and from type 2 back to type 2. This is because the angle constraint on the input polyhedron prevents parent–child relations for points on segments and facets with a nonempty intersection. If there is a relation between points on segments and facets with an empty intersection, then $r_x \geq f(x)$ and there is no need to follow an arc in the digraph.

Observe that every cycle in the digraph contains the arc labeled ϱ_0 leading into type 3. We choose $\varrho_0 \geq 2$ to guarantee that the products of arc labels for all cycles are 1 or larger. The smallest product of any path in the digraph is therefore $1/2$. In cases where r_x is not at least $f(x)$, there exist ancestors q with $r_x \geq r_q/2$ and $r_q \geq f(q)$. Since $f(q)$ is bounded away from zero by some positive constant, we conclude that the insertion radii cannot get arbitrarily small. It follows that the Delaunay refinement algorithm terminates. For $\varrho_0 < 2$ there are cases where the algorithm does not terminate.

Graded meshes

With little additional effort we can show that for ϱ_0 strictly larger than 2, insertion radii are directly related to local feature size, and not just indirectly through chains of ancestors. We begin with a relation between the local feature size over insertion radius ratio of a vertex and of its parent.

Ratio Claim. Let x be a Delaunay vertex with parent p and assume $r_x \geq c \cdot r_p$. Then

$$f(x)/r_x \leq 1 + f(p)/(c \cdot r_p).$$

Proof. We have $r_x = \|x - p\|$ if p is a Delaunay vertex and $r_x \geq \|x - p\|$ if p is a rejected midpoint or circumcenter. Starting with the Lipschitz condition, we get

$$f(x) \leq f(p) + \|x - p\|$$
$$\leq (f(p)/c \cdot r_p) \cdot r_x + r_x,$$

and the result follows after dividing by r_x.   □

To prepare the next step we assume $\varrho_0 > 2$ and define constants

$$C_1 = \frac{(3 + \sqrt{2}) \cdot \varrho_0}{\varrho_0 - 2},$$

$$C_2 = \frac{(1 + \sqrt{2}) \cdot \varrho_0 + \sqrt{2}}{\varrho_0 - 2},$$

$$C_3 = \frac{\varrho_0 + 1 + \sqrt{2}}{\varrho_0 - 2}.$$

Note that $C_1 > C_2 > C_3 > 1$.

Invariant. If x is a type i vertex in the Delaunay tetrahedrization, for $1 \leq i \leq 3$, then $r_x \geq f(x)/C_i$.

Proof. If the parent p of x is a vertex of the input polyhedron P then $r_x \geq f(x)$ and we are done. Otherwise, assume inductively that the claimed inequality holds for vertex p. We finish the proof by case analysis. If x has type 3 then $c = \varrho_0$ and $r_x \geq \varrho_0 \cdot r_p$ by the Radius Claim. By induction we get $f(p) \leq C_1 r_p$, no matter what type p has. Using the Ratio Claim we get

$$f(x)/r_x \leq 1 + C_1/\varrho_0 = C_3.$$

If x has type 2 then $c = 1/\sqrt{2}$. We have $r_x \geq f(x)$ unless p has type 3, and therefore $f(p) \leq C_3 r_p$ by inductive assumption. Then $r_x \geq r_p/\sqrt{2}$ by the Radius Claim, and

$$f(x)/r_x \leq 1 + \sqrt{2} \cdot C_3 = C_2$$

by the Ratio Claim. If x has type 1 then $c = 1/\sqrt{2}$. We have $r_x \geq f(x)$ unless p has type 2 or 3, and therefore $f(p) \leq C_2 r_p$ by inductive assumption. Then $r_x \geq r_p/\sqrt{2}$ by the Radius Claim, and

$$f(x)/r_x \leq 1 + \sqrt{2} \cdot C_2 = C_1$$

by the Ratio Claim. □

Because C_1 is the largest of the three constants, we can simplify the Invariant to $r_x \geq f(x)/C_1$ for every Delaunay vertex x. From this we conclude

$$\|x - y\| \geq f(x)/(1 + C_1)$$

for any two vertices x, y in the Delaunay tetrahedrization, using the argument in the proof of the Smallest Gap Lemma in Section 2.3.

Bibliographic notes

The bulk of the material in this section is taken from a paper by Jonathan Shewchuk [2]. In that paper, the assumed input is a so-called piecewise linear complex as defined by Miller et al. [1]. This is a 3-face of a polyhedron together with its faces, which is slightly more general than a three-dimensional polyhedron.

[1] G. L. Miller, D. Talmor, S.-H. Teng, N. Walkington, and H. Wang. Control volume meshes using sphere packing: generation, refinement and coarsening. In "Proc. 5th Internat. Meshing Roundtable," 1996, 47–61.
[2] J. R. Shewchuk. Tetrahedral mesh generation by Delaunay refinement. In "Proc. 14th Ann. Sympos. Comput. Geom.," 1998, 86–95.

6.4 Sliver exudation

The sliver is the only type of small volume tetrahedron whose circumradius over the shortest edge length ratio does not grow with decreasing volume. Experimental studies indicate that slivers frequently exist right between other well-shaped tetrahedra inside Delaunay tetrahedrizations. This section explains how point weights can be used to remove slivers.

Periodic meshes

Suppose S is a finite set of points in $\mathbb{R}^3$ whose Delaunay tetrahedrization has
the ratio property for a constant ρ_0. The goal is to prove that there are weights
we can assign to the points such that the weighted Delaunay tetrahedrization is
free of slivers. This cannot be true in full generality, for if S consists of only
four points forming a sliver then no weight assignment can make that sliver
disappear. We avoid this and similar boundary effects by replacing the finite by
a periodic set $S = P + \mathbb{Z}^3$, where P is a finite set of points in the half-open
unit cube $[0, 1)^3$ and $\mathbb{Z}^3$ is the three-dimensional integer grid. The periodic set
S contains all points $p + \mathbf{v}$, where $p \in P$ and $\mathbf{v}$ is an integer vector. Like S, the
Delaunay tetrahedrization D of S is periodic. Specifically, for every tetrahedron
$\tau \in D$, the shifted copies $\tau + \mathbb{Z}^3$ are also in D. This idea is illustrated for a periodic set generated by three points in the half-open unit square in Figure 6.12.

Weight assignment

A *weight assignment* is a function $\omega : P \to \mathbb{R}$. The resulting set of spheres is
denoted as $S_\omega = \{(a, \omega(p)) \mid p \in P, a \in p + \mathbb{Z}^3\}$. Depending on ω, a point p
may or may not be a vertex of the weighted Delaunay triangulation of S_ω, which
we denote as D_ω. Let $N(p)$ be the minimum distance to any other point in S.
To prevent points from becoming redundant, we limit ourselves to *mild* weight
assignments that satisfy $0 \le \omega(p) \le N(p)/3$ for all $p \in P$. Every sphere in

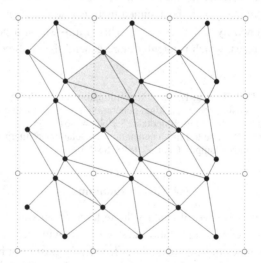

Figure 6.12. Periodic tiling of the plane. The shaded triangles form a domain whose
shifted copies tile the entire plane.

S_ω has a real radius and every pair is disjoint and not nested. It follows that none of the points is redundant. Another benefit of a mild weight assignment is that it does not drastically change the shape of triangles and tetrahedra. In particular, D_ω has the ratio property for a constant ϱ_1 that only depends on ϱ_0. It follows that the area of each triangle is bounded from below by some constant times the square of its circumcircle. The same is not true for volumes of tetrahedra, which is why eliminating slivers is difficult.

A crucial step toward eliminating slivers is a generalization of the Degree Lemma of Section 6.2. Let K be the set of simplices that occur in weighted Delaunay tetrahedrizations for mild weight assignments of S. In other words, $K = \bigcup_\omega D_\omega$, which is a three-dimensional simplicial complex but not necessarily geometrically realized in $\mathbb{R}^3$. The vertex set of K is Vert $K = S$, and the *degree* of a vertex is the number of edges in K that share the vertex.

Weighted Degree Lemma. There exists a constant δ_1 depending only on ϱ_0 such that the degree of every vertex in K is at most δ_1.

The proof is fairly tedious and partially a repeat of the proofs of the Length Variation and Degree Lemmas of Section 6.2. It is therefore omitted.

Slicing orthogonal spheres

We need an elementary fact about spheres (a, A^2) and (z, Z^2) that are orthogonal; that is, $\|a - z\|^2 = A^2 + Z^2$. A plane intersects the two spheres in two circles, which may have real or imaginary radii.

Slicing Lemma. A plane passing through a intersects the two spheres in two orthogonal circles.

Proof. Let (x, X^2), (y, Y^2) be the circles where the plane intersects the two spheres. We have $x = a$, $X^2 = A^2$, and $Y^2 = Z^2 - \|z - y\|^2$. Hence

$$\|x - y\|^2 = \|x - z\|^2 - \|z - y\|^2$$
$$= (A^2 + Z^2) - (Z^2 - Y^2)$$
$$= X^2 + Y^2.$$

In words, the two circles are also orthogonal. $\qquad\square$

As an application of the Slicing Lemma consider three spheres and the plane that passes through their centers, as in Figure 6.13. The plane intersects the

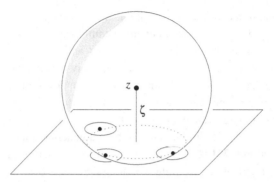

Figure 6.13. Slice through three spheres and another sphere orthogonal to the first three.

three spheres in three circles, and there is a unique circle orthogonal to all three. The Slicing Lemma implies that every sphere orthogonal to all three spheres intersects the plane in this same circle.

Variation of orthoradius

Another crucial step toward eliminating slivers is the stability analysis of their orthospheres. We will see that a small weight change can increase the size of the orthosphere dramatically. This is useful because a tetrahedron in D_ω cannot have a large orthosphere, for else that orthosphere would be closer than orthogonal to some weighted point. We later exploit this observation and change weights to increase orthospheres of slivers.

Let us analyze how the radius of the orthosphere of four spheres changes as we manipulate the weight of one of the sphere. Let (y, Y^2) be the smallest sphere orthogonal to the first three spheres, let (p, P^2) be the fourth sphere, and let (z, Z^2) be the orthosphere of all four spheres, as illustrated in Figure 6.14.

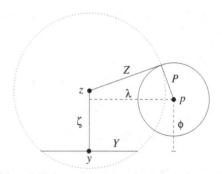

Figure 6.14. The orthocenter z moves downward as the weight of p increases.

Let ζ and ϕ be the distances of z and p from the plane h that passes through the centers of the first three spheres. With varying P^2, the center of the orthosphere moves along the line that meets h orthogonally at y. The distance of z from h is a function of the weight of p, $\zeta : \mathbb{R} \rightarrow \mathbb{R}$.

Distance Variation Lemma. $\zeta(P^2) = \zeta(0) - P^2/2\phi$.

Proof. Let λ be the distance from p to the line along which z moves. We have $Z^2 + P^2 = (\zeta(P^2) - \phi)^2 + \lambda^2$. The weight of the orthosphere is $Z^2 = \zeta(P^2)^2 + Y^2$. Hence

$$\zeta(P^2)^2 = Z^2 - Y^2$$
$$= (\zeta(P^2) - \phi)^2 + \lambda^2 - P^2 - Y^2.$$

After canceling $\zeta(P^2)^2$, we get

$$\zeta(P^2) = (\phi^2 + \lambda^2 - Y^2)/2\phi - P^2/2\phi.$$

The first term on the right-hand side is $\zeta(0)$. $\qquad\qquad\qquad\qquad\qquad$ □

The term $P^2/2\phi$ is the displacement of the orthocenter that occurs as we change the weight of p from 0 to P^2. For slivers, the value of ϕ is small, which implies that the displacement is large.

Sliver theorem

We finally show that there is a mild weight assignment that removes all slivers. The proof is constructive and assigns weights in sequence to the points in P. To quantify the property of being a sliver, we define $\xi = V/L$, where V is the volume and L is the length of the shortest edge of the tetrahedron. Only slivers can have bounded R/L as well as small ξ. Note that the volume of the tetrahedron indicated in Figure 6.14 is one-third the area of the base triangle times ϕ. As mentioned above, the area of the base triangle is some positive constant fraction of Y^2. Similarly, L is some positive constant fraction of Y, which implies that ξ is some positive constant fraction of $Y\phi$.

Sliver Theorem. There are constants $\varrho_1, \xi_0 > 0$ and a mild weight assignment ω, such that the weighted Delaunay tetrahedrization has the ratio property for ϱ_1 and $\xi > \xi_0$ for all its tetrahedra.

Proof. We focus on proving $\xi > \xi_0$ for all tetrahedra in D_ω. Assume without loss of generality that the distance from a point p to its nearest neighbor in S

is $N(p) = 1$. The weight assigned to p can be anywhere in the interval $[0, \frac{1}{3}]$. According to the Weighted Degree Lemma, there is only a constant number of tetrahedra that can possibly be in the star of p. Each such tetrahedron can exist in D_ω only if its orthosphere is not too big. In other words, the tetrahedron can only exist if $\omega(p)$ is chosen inside some subinterval of $[0, \frac{1}{3}]$. The Distance Variation Lemma implies that the length of this subinterval decreases linearly with ϕ and therefore linearly with ξ. We can choose ξ_0 small enough such that the constant number of subintervals cannot possibly cover $[0, \frac{1}{3}]$. By the pigeonhole principle, there is a value $\omega(p) \in [0, \frac{1}{3}]$ that excludes all slivers from the star of p.   □

Removing slivers

The proof of the Sliver Theorem suggests an algorithm that assigns weights to individual points in an arbitrary sequence. For each point $p \in P$, the algorithm considers the interval of possible weights and the subintervals in which tetrahedra in K can occur in the weighted Delaunay tetrahedrization. We could consider all tetrahedra in the star of p in K, but it is more convenient to consider only the subset in the 1-parameter family of weighted Delaunay tetrahedrizations generated by continuously increasing the weight of p from 0 to $N(p)/3$. For each such tetrahedron, we get the ξ value and a subinterval during which it exists in D_ω. Figure 6.15 draws each tetrahedron as a horizontal line segment in the $\omega\xi$-plane. The lower envelope of the line segments is the function that maps the weight of p to the worst ξ value of any tetrahedron in its star. The algorithm finds the weight where that function has a maximum and assigns it to p. Since there is only a constant number of tetrahedra to be considered, this can be accomplished in constant time. The overall running time of the algorithm is therefore $O(n)$, where $n = \text{card } P$.

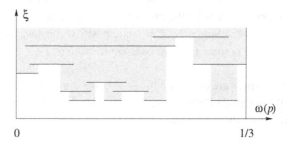

Figure 6.15. Each tetrahedron in the star is represented by a horizontal line segment.

A source of possible worry is that, after we have fixed the weight of p, we may modify the weight of a neighbor q of p. Modifying the weight of q may change the star of p. However, all new tetrahedra in the star of p also belong to the star of q and thus cannot have arbitrarily small ξ values. We thus do not have to reconsider p, and $O(n)$ time indeed suffices. The Sliver Theorem guarantees the algorithm is successful as quantified by the positive constant ξ_0. While the algorithm does not find the globally optimum weight assignment, it finds the optimum for each point individually assuming fixed weights of other points. It might therefore achieve a minimum ξ value that is much better than the rather pessimistic estimate for ξ_0 guaranteed by the Sliver Theorem.

Bibliographic notes

The material of this section is taken from the sliver exudation paper by Cheng et al. [2]. The occurrence of slivers as a menace in three-dimensional Delaunay tetrahedrizations was reported by Cavendish, Field and Frey [1]. Besides the sliver exudation method described in this section, there are two other methods that provably remove slivers. The first, by Chew [3], adds points and uses randomness to avoid creating new slivers. The second, by Edelsbrunner et al. [4], moves points and relies on the ratio property of the Delaunay tetrahedrization, as in the weight assignment method of this section.

[1] J. C. Cavendish, D. A. Field, and W. H. Frey. An approach to automatic three-dimensional finite element mesh generation. *Internat. J. Numer. Methods Engrg.* **21** (1985), 329–347.
[2] S.-W. Cheng, T. K. Dey, H. Edelsbrunner, M. A. Facello, and S.-H. Teng. Sliver exudation. *J. Assoc. Comput. Mach.* **47** (2000), 883–904.
[3] L. P. Chew. Guaranteed-quality Delaunay meshing in 3D. Short paper in "Proc. 13th Ann. Sympos. Comput. Geom.," 1997, 391–393.
[4] H. Edelsbrunner, X.-Y. Li, G. L. Miller, A. Stathopoulos, D. Talmor, S.-H. Teng, A. Üngör, and N. Walkington. Smoothing cleans up slivers. In "Proc. 32nd Ann. ACM Sympos. Theory Comput.," 2000, 273–277.

Exercise collection

The credit assignment reflects a subjective assessment of difficulty. A typical question can be answered by using knowledge of the material combined with some thought and analysis.

1. **Removing vertices** (two credits). Let P be a convex polytope with n vertices in $\mathbb{R}^3$. Tetrahedrize P by selecting a vertex u, reducing P to the convex hull of the remaining vertices, and forming tetrahedra as cones from u to new

triangles in the boundary. This step is repeated until P is a tetrahedron. Prove that there is an ordering of the vertices such that the algorithm constructs at most $3n - 11$ tetrahedra.

2. **Interior edges** (two credits). Let P be a convex polytope with n vertices in $\mathbb{R}^3$ and K a tetrahedrization whose only vertices are the ones of P. An *interior edge* of K passes through the interior of P.

   (i) Show if K contains no interior edge then the number of tetrahedra is $n - 3$.

   (ii) What is the number of tetrahedra if K contains t interior edges?

3. **Tetrahedrizing the cube** (two credits). Consider the unit cube, $[0, 1]^3$, and let K be a tetrahedrization whose only vertices are the eight corners of the cube.

   (i) Prove that K either contains five or six tetrahedra.

   (ii) Draw all nonisomorphic such tetrahedrizations K of the cube, and their dual graphs.

4. **BCC tetrahedron** (one credit). The *body-centered cube (BCC) lattice* consists of the integer points (i, j, k), where all three coordinates are either even or all three are odd. All Delaunay tetrahedra of the BCC lattice are congruent to a single tetrahedron, which we call the *BCC tetrahedron*. What is the circumradius to the shortest edge length ratio of that tetrahedron?

5. **Packing** (three credits). A collection of closed unit balls in $\mathbb{R}^3$ forms a *packing* if their interiors are pairwise disjoint. The packing is *maximal* if no unit ball can be added without overlapping the interior of other balls. Let $S \subseteq \mathbb{R}^3$ such that the set of unit balls $S + \mathbb{B}^3$ is a maximal packing.

   (i) Show that increasing the balls to twice the size produces a covering of space; that is,

$$\bigcup (S + 2\mathbb{B}^3) = \mathbb{R}^3.$$

   (ii) Prove that an edge in the Delaunay tetrahedrization of S has length at most 4.

   (iii) Prove that there is a constant c such that each vertex in the Delaunay tetrahedrization of S belongs to at most c edges.

6. **Faces of a polyhedron** (one credit). Count the segments and facets of the polyhedron in Figure 6.16 by using the definition in Section 6.1. How many of the segments and facets consist of more than one connected component?

7. **Angles of a tetrahedron** (three credits). Let $abcd$ be a tetrahedron in $\mathbb{R}^3$, let $\mathbb{S}^2$ be the unit sphere centered at the origin, and let $\varepsilon > 0$ be sufficiently small. Recall that 4π is the area of $\mathbb{S}^2$.

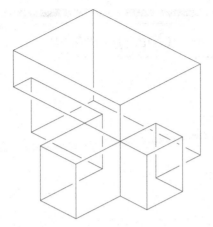

Figure 6.16. A nonconvex polyhedron.

(i) The *solid angle* at a is 4π times the fraction of $a + \varepsilon\mathbb{S}^2$ inside the tetrahedron. Prove that the sum of solid angles at a, b, c, d is a real number between zero and 2π.

(ii) The *dihedral angle* at ab is 2π times the fraction of $x + \varepsilon\mathbb{S}^2$ inside $abcd$, where x is an interior point of ab. Prove that the sum of dihedral angles at ab, ac, ad, bc, bd, cd is between 2π and 3π.

(iii) Show that twice the sum of six dihedral angles exceeds the sum of four solid angles by 4π.

7

Open problems

This chapter collects open problems that in one way or the other relate to the material discussed in this book. They represent the complement of the material, in the sense that they attempt to describe what we do not know. We should keep in mind that it is most likely the case that only a tiny fraction of the knowable is known. Hence, there is a vast variety of questions that can be asked but not yet answered. The author of this book exercised subjective taste and judgment to collect a small subset of such questions, in the hope that they can give a glimpse of what is conceivable. Most of the problems are elementary in nature and have been stated elsewhere in the literature.

Two of the twenty-three problems have been solved since this book first appeared in 2001. These are **P.8 Union of disks** and **P.9 Intersection of disks**, both solved in [1]. Since the described approaches to the two problems are different from the eventual solution and perhaps useful to understand generalized versions of the problem, we decided to leave Sections P.8 and P.9 unchanged.

[1] K. Bezdek and R. Connelly. Pushing disks apart—the Kneser-Poulsen conjecture in the plane. *J. Reine Angew. Math.* **553** (2002), 221–236.

P.1 Empty convex hexagons

Let S be a set of n points in $\mathbb{R}^2$ and assume no three points are collinear. A *convex k-gon* is a subset of k points in convex position. We call a convex k-gon *empty* if every point of S either belongs to the subset or lies outside the convex hull of the subset; see Figure 7.1. Erdős and Szekeres proved in 1935 that there exists n_k such that card $S \geq n_k$ implies that S contains at least one convex k-gon [1]. Their lower bound is $2^{k-2} + 1 \leq n_k$ and this is conjectured to be tight. Their upper bound has only been improved marginally, and the current best bound is $n_k \leq \binom{2k-5}{k-2} + 2$, proved by Tóth and Valtr [5].

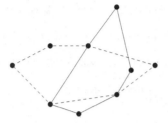

Figure 7.1. There are nine points; the dashed edges show a convex 6-gon that is nonempty, and the solid edges show a convex 6-gon that is empty.

Let m_k be the corresponding number that guarantees the existence of an empty convex k-gon. We have $m_3 = 3$ and $m_4 = 5$. A version of the following argument bounding m_5 appears in a survey paper by Paul Erdős and is attributed to Andrzej Ehrenfeucht.

Claim. $m_5 \leq 37$.

Proof. Let S be a set of at least 37 points. Since $n_6 \leq 37 = \binom{2 \cdot 6 - 5}{6 - 2} + 2$, there exist convex 6-gons. Take the one with fewest points inside, and let this number be i. If $i = 0$ we have an empty convex 6-gon and are done. If $i = 1, 2$ we can find an empty convex 5-gon directly, as shown in Figure 7.2(a) and 7.2(b). If $i \geq 3$ we take the convex hull of the i points and consider the line defined by one of the convex hull edges. Either there is an empty convex 5-gon on the other side of that line, as in Figure 7.2(c), or we get another convex 6-gon with only $i - 2$ points inside, as in Figure 7.2(d), which contradicts the minimality assumption. $\square$

Heiko Harborth proves $m_5 = 10$, which is larger than $n_5 = 9$ [2]. Joe Horton proves that m_7 does not exist [3]. Overmars, Scholten, and Vincent use the computer to construct a set of 28 points without an empty convex hexagon [4].

Question. Does m_6 exist?

[1] P. Erdős and G. Szekeres. A combinatorial problem in geometry. *Compositio Math.* **2** (1935), 463–470.

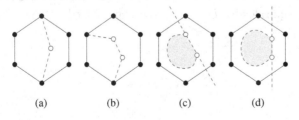

      (a)   (b)   (c)   (d)

Figure 7.2. Different configurations of i points inside a convex hexagon.

[2] H. Harborth. Konvexe Fünfecke in ebenen Punktmengen. *Elem. Math.* **33** (1978), 116–118.

[3] J. D. Horton. Sets with no empty convex 7-gon. *Canad. Math. Bull.* **26** (1983), 482–484.

[4] M. H. Overmars, B. Scholten, and I. Vincent. Sets without empty convex 6-gons. *Bulletin EATCS* **37** (1989), 160.

[5] G. Tóth and P. Valtr. Note on the Erdős-Szekeres theorem. *Discrete Comput. Geom.* **19** (1998), 457–495.

P.2 Unit distances in the plane

Let S be a set of n points in $\mathbb{R}^2$. How many of the $\binom{n}{2}$ pairs can be exactly one unit of distance apart? To state partial answers, let $f(S)$ be the number of unit-distance pairs and define

$$f(n) = \max\{f(S) \mid S \subseteq \mathbb{R}^2, \text{card } S = n\}.$$

Paul Erdős [2] studied this question in a paper published in 1946, where he proved there exist constants c and c' such that

$$n^{1+\frac{c}{\log_2 \log_2 n}} \leq f(n) \leq c' \cdot n^{3/2}.$$

The geometric construction for the lower bound is simple, namely a square grid of points in the plane. The analysis of this example, however, is involved and requires nontrivial number theoretic results. To prove the upper bound, consider the graph whose vertices are the points in S and whose edges all have unit length. Since two circles meet in at most two points, this graph contains no complete bipartite subgraph of two plus three vertices; see Figure 7.3. Assume vertex p_i has d_i unit distance neighbors. There are $\binom{d_i}{2}$ pairs, and if we count neighbor pairs over all $p_i \in S$ then each pair is counted at most twice. Hence

$$\sum_{p_i \in S} \binom{d_i}{2} \leq 2\binom{n}{2}.$$

To maximize $\sum d_i$ we may assume that all d_i are about the same, namely $\binom{d_i}{2} \approx 2\binom{n}{2}/n$, or equivalently, $d_i \approx \sqrt{2n}$. The upper bound follows. The $c' \cdot n^{3/2}$

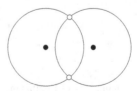

Figure 7.3. At most two points can have distance 1 from two other points.

bound has since been improved to constant times $n^{4/3}$ by Spencer, Szemerédi, and Trotter [3].

Question. Is it true that for every $\varepsilon > 0$ there is a constant $c = c(\varepsilon)$ such that $f(n) \leq c \cdot n^{1+\varepsilon}$?

Two points a and b form a unit distance pair if and only if a lies on the unit circle around b, and conversely b lies on the unit circle around a. We can therefore think of the unit distance problem as a special case of counting the incidences between n points and n unit circles. What happens if we drop the constraint that all circles be the same size? Then the lower bound goes up to constant times $n^{4/3}$, and the current best upper bound given by Clarkson et al. [1] is constant times $n^{7/5}$.

[1] K. L. Clarkson, H. Edelsbrunner, L. J. Guibas, M. Sharir, and E. Welzl. Combinatorial complexity bounds for arrangements of curves and spheres. *Discrete Comput. Geom.* **5** (1990), 99–160.
[2] P. Erdős. On sets of distances of n points. *Amer. Math. Monthly* **53** (1946), 248–250.
[3] J. Spencer, E. Szemerédi, and W. Trotter, Jr. Unit distances in the Euclidean plane. In *Graph Theory and Combinatorics*, B. Bollobás (ed.), Academic Press, New York, 1984, 293–303.

P.3 Convex unit distances

Let S be a set of n points in the plane. Assume the points are in *convex position*, by which we mean that they are the vertices of a convex polygon. We disallow collinear vertex triplets by requiring that the angle at each vertex be strictly less than π. Let $u(S)$ be the number of point pairs at unit distance,

$$u(S) = \text{card}\left\{\{x, y\} \in \binom{S}{2} \mid \|x - y\| = 1\right\},$$

and let $u(n)$ be the maximum number of unit distance pairs over all sets $S \subseteq \mathbb{R}^2$ of n points in convex position.

The problem of determining $u(n)$ was stated in a paper by Erdős and Moser [2] together with a lower bound of roughly $5n/3 \leq u(n)$. The currently best lower bound of $2n - 7$ given by Edelsbrunner and Hajnal [1] is illustrated in Figure 7.4. Start the construction with an equilateral Reuleaux triangle ABC. Points A, B, C are auxiliary points of the construction. Let a, b, c be the midpoints of the circular arcs. Choose a point a_1 at unit distance from a and use it as the starting point of a chain $a_1, b_1, c_1, a_2, b_2, c_2, a_3$, and so on. Consecutive points in this chain are at unit distance from each other, and also $\|a - a_i\| = \|b - b_i\| = \|c - c_i\| = 1$ for every i. The chain contains $n - 4$ unit distance pairs and we get an additional $n - 3$ pairs from a, b, c to points in the

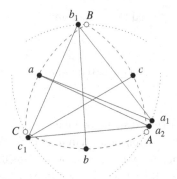

Figure 7.4. The solid edges have unit length and connect a to the a_i, b to the b_i, c to the c_i, and the indexed points in a sequence.

chain. The construction works because the a_i monotonically approach a from one side as i goes to infinity, and similarly for the b_i and the c_i.

To get an upper bound, note that at most two points can be at unit distance from two given points. The graph of unit distance pairs thus contains no complete bipartite subgraph of 2 plus 3 nodes. In other words, the adjacency matrix contains no two-by-three submatrix full of ones. Using the convexity condition, Zoltan Füredi [3] further shows that the adjacency matrix contains no submatrix of the form

$$\begin{pmatrix} 1 & 1 & * \\ 1 & * & 1 \end{pmatrix},$$

where $*$ can be either 0 or 1. He proves that every such matrix has at most some constant time $n \log_2 n$ ones, which implies the same upper bound for $u(n)$.

Question. Is there a constant c such that $u(n) \leq c \cdot n$?

[1] H. Edelsbrunner and P. Hajnal. A lower bound on the number of unit distances between the vertices of a convex polygon. *J. Combin. Theory, Ser. A* **56** (1991), 312–316.
[2] P. Erdős and L. Moser. Problem 11. *Canad. Math. Bull.* **2** (1959), 43.
[3] Z. Füredi. The maximum number of unit distances in a convex n-gon. *J. Combin. Theory, Ser. A* **55** (1990), 316–320.

P.4 Bichromatic minimum distances

Let W be a set of n white and B a set of n black points. Assume the minimum distance between a white and a black point is 1. A *bichromatic minimum distance pair* is an edge $wb \in W \times B$ of length $\|w - b\| = 1$. Let $\beta(W, B)$ be the number of bichromatic minimum distance pairs, and let $\beta_d(n)$ be the maximum over all sets W and B of n points in $\mathbb{R}^d$ each.

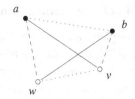

Figure 7.5. Diagonals and edges of a convex quadrangle. The shorter dashed edge is shorter than the longer diagonal.

In $\mathbb{R}^2$ we have $\beta_2(n) \leq 4n - 4$. To prove the upper bound, note first that the edges we count are pairwise noncrossing. Indeed, if wb and va cross then

$$\min\{\|w - a\|, \|v - b\|\} < 1,$$

as illustrated in Figure 7.5. This contradicts the assumption that one is the distance between W and B. The graph whose vertices are the $2n$ points and whose edges are the bichromatic minimum distance pairs is therefore planar. The graph is also bipartite, which implies that each face has an even number of edges. We can add edges until the graph is connected and each face is a quadrangle. The Euler relation for such a graph is $2n - e + f = 2$, where e is the number of edges and f is the number of faces. We combine this with $4f = 2e$ and get $e = 4n - 4$.

In $\mathbb{R}^4$ we can choose n white points on the circle $x_1^2 + x_2^2 = 1, x_3 = x_4 = 0$, and n black points on the circle $x_1 = x_2 = 0, x_3^2 + x_4^2 = 1$. Then $\|w - b\| = 1$ for every $wb \in W \times B$. This implies $\beta_d(n) = n^2$ for all $d \geq 4$.

The only difficult case is in $\mathbb{R}^3$ where the current best upper bound is a constant times $n^{4/3}$ proved by Edelsbrunner and Sharir [1]. No superlinear lower bound is known.

Question. Is there a constant c such that $\beta_3(n) \leq c \cdot n$?

For the monochromatic case we can use a packing argument to show that such a constant exists. The problem is the same as arranging n equal spheres in $\mathbb{R}^3$ so that no two overlap and we have as many touching pairs as possible. A single sphere cannot touch more than 12 others, which implies that the number of touching pairs is at most $6n$. The packing argument fails in the bichromatic case because points of the same color can be arbitrarily close to each other.

[1] H. Edelsbrunner and M. Sharir. A hyperplane incidence problem with applications to counting distances. In *Applied Geometry and Discrete Mathematics. The Victor Klee Festschrift*, P. Gritzmann and B. Sturmfels (eds.), AMS and ACM, 1991, 253–263.

P.5 MinMax area triangulation

Let S be a set of n points in the plane, and consider the collection of all possible triangulations of S. Among these, the Delaunay triangulation maximizes the smallest angle, and it minimizes the largest circumcircle. Rajan [3] proves that the Delaunay triangulation also minimizes the largest minidisk, which for a triangle is the smallest disk that contains it.

The general problem of computing an optimal triangulation under some measure is, however, difficult. There are usually exponentially many triangulations, and enumerating all is not practical, unless n is very small. A class of optimization problems with polynomial time algorithms has been identified by Bern, et al. [1]. They generalize the $O(n^2 \log n)$ time algorithm for minimizing the maximum angle given in [2] and formulate an abstract condition for measures under which the algorithm succeeds to find the optimum. We need definitions to explain the condition.

A *measure* maps a triangle to a real number. We consider MinMax problems, in which the measure of a triangulation K is $\mu(K) = \max\{\mu(xyz) \mid xyz \in K\}$. The *worst* triangles in K are the ones with measure equal to $\mu(K)$. A triangulation K of S *breaks* a triangle $xyz \in \binom{S}{3}$ *at* y if it contains an edge yt that crosses xz. Vertex y is *anchor* of xyz if every triangulation K with $\mu(K) \leq \mu(xyz)$ either contains xyz or breaks xyz at y.

Anchor Condition. The worst triangles of every triangulation have anchors.

If μ satisfies the Anchor Condition, and if the anchor of a triangle can be computed in constant time, then the algorithm in [1] constructs an optimum triangulation of S in time $O(n^3)$. If all triangles in every triangulation have anchors, and not just the worst ones, then the running time can be improved to $O(n^2 \log n)$.

The maximum angle inside a triangle obviously satisfies the stronger form of the Anchor Condition. The distance between a triangle and its circumcenter satisfies the Anchor Condition, but not its stronger form, and thus can be minimized in time $O(n^3)$. The negative height of a triangle satisfies the stronger form of the Anchor Condition, so height can be maximized in time $O(n^2 \log n)$. A quality measure that does not satisfy the Anchor Condition is triangle area.

Question. Is there a polynomial time algorithm for computing a triangulation that minimizes the largest triangle area?

Maybe the Anchor Condition can be relaxed to cover the area measure without sacrificing the polynomial time bound. The MinMax area question has a

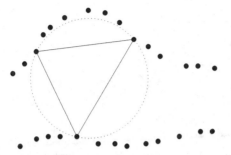

Figure 7.6. The Delaunay triangle has large area and does not belong to the MinMax area triangulation.

practical motivation. Points often come with measurements, which can be heights within a landscape, depths within a river, and so on. In most land or water surveys, the points are nowhere near a random distribution but rather reflect characteristic patterns implied by the data collection mechanism. For example, if the depth of a river is measured from two boats, we are likely to get two wavy lines of points such as the ones in Figure 7.6. The triangulation is used to extend the measurements to a piecewise linear function over the convex hull. The measurements usually have errors, and the goal is to avoid spreading the error of any one measurement over a large region.

[1] M. Bern, H. Edelsbrunner, D. Eppstein, S. Mitchell, and T. S. Tan. Edge insertion for optimal triangulations. *Discrete Comput. Geom.* **10** (1993), 47–65.
[2] H. Edelsbrunner, T. S. Tan, and R. Waupotitsch. An O($n^2 \log n$) time algorithm for the MinMax angle triangulation. *SIAM J. Statist. Sci. Comput.* **13** (1992), 66–104.
[3] V. T. Rajan. Optimality of the Delaunay triangulation in $\mathbb{R}^d$. *Discrete Comput. Geom.* **12** (1994), 189–202.

P.6 Counting triangulations

Let S be a set of n points in the plane. By a triangulation of S we mean as usual an edge-to-edge decomposition of the convex hull into triangles whose vertices are the points in S. If the n points are in convex position, then the number of different ways to triangulate S is

$$t(S) = \binom{2n-4}{n-2} \Big/ (n-1),$$

which is at most 2^{n-3}. There is an elegant argument that establishes this equation.

A triangulation of a convex $(k+2)$-gon has a dual tree with k interior nodes, each of degree 3. We define a root by removing one edge of the $(k+2)$-gon,

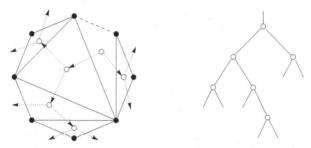

Figure 7.7. Triangulation of a convex $(k + 2)$-gon, for $k + 2 = 8$, and the dual binary tree with k interior nodes. The corresponding well-formed string is $LLLRRLRLR$ RLR.

as illustrated in Figure 7.7. Orient the edges away from the root and use the layout to distinguish between left and right outgoing edges. We traverse the tree, always first visiting the left and then the right subtree. The traversal defines a string, where L records a left edge down and R records a right edge down. There is one left and one right edge per interior node, which implies the string consists of k Ls and k Rs. Because the left edge of each node precedes its right edge, each prefix of the string contains at least as many Ls as Rs. We call such a string *well formed* and note that there is a bijection between binary trees with k interior nodes and well-formed strings of length $2k$.

Claim. The number of well formed strings of length $2k$ is

$$\binom{2k}{k} \Big/ (k + 1).$$

Proof. The total number of strings formed by k Ls and k Rs is $\binom{2k}{k}$. If a string is not well formed, we invert the smallest prefix that has more Rs than Ls. For example, we change $LLRRR\text{-}LLR$ to $RRLLL\text{-}LLR$. The new string has $k + 1$ Ls and $k - 1$ Rs, and the operation is reversible. So there are $\binom{2k}{k-1}$ strings of k Ls and k Rs that are not well formed.

$$\binom{2k}{k} - \binom{2k}{k-1} = \left(1 - \frac{k}{k+1}\right) \cdot \binom{2k}{k}$$

$$= \frac{1}{k+1} \cdot \binom{2k}{k}.$$

$\square$

The claimed number of triangulations follows by setting $n = k - 2$. In general, the number of triangulations does not only depend on n but also on S. It is known that the number is at most $t(S) \le c^n$, for some constant $c > 0$. This bound is a shared consequence of different combinatorial results by Tutte [3] and by Ajtai et al. [2]. Even to compute a random triangulations of a given set S seems difficult. Related to picking a random element is counting the possibilities.

Question. Is there a polynomial time algorithm for counting the triangulations of a set of n points in the plane?

An algorithm that counts in time sublinear in the number of triangulations has recently been found by Aichholzer [1].

[1] O. Aichholzer. The path of a triangulation. In "Proc. 15th Ann. Sympos. Comput. Geom.," 1999, 14–23.
[2] M. Ajtai, V. Chvátal, M. M. Newborn, and E. Szemerédi. Crossing-free subgraphs. *Ann. Discrete Math.* **12** (1982), 9–12.
[3] W. T. Tutte. A census of planar triangulations. *Canad. J. Math.* (1962), 21–38.

P.7 Sorting $X + Y$

Let M be an n-by-n matrix of real numbers. We can sort the n^2 numbers in $O(n^2 \log n)$ time by using heapsort or any one of a number of other asymptotically optimal sorting algorithms. Since there are $n^2!$ possible permutations, every comparison-based algorithm takes at least $\log_2 n^2!$ or about $2n^2 \log_2 n$ comparisons and time. As shown in [3], the lower bound even applies if the rows and columns of M are already sorted.

Now suppose that $X = (x_1, x_2, \ldots, x_n)$ and $Y = (y_1, y_2, \ldots, y_n)$ are two vectors each, and $M = X + Y$. By this we mean that the element in the ith row and the jth column is $m_{ij} = x_i + y_j$, as illustrated in Figure 7.8. We may

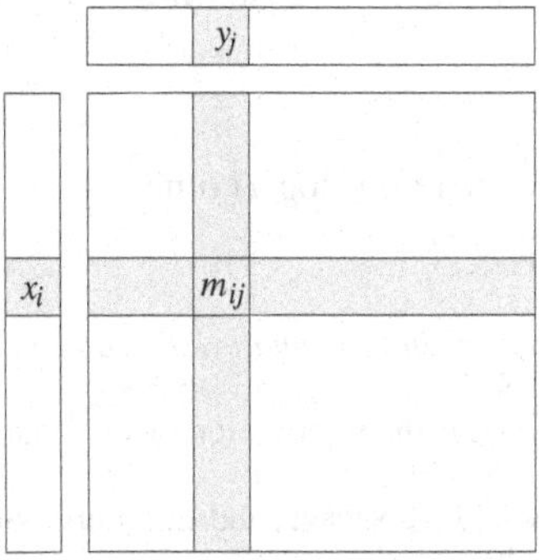

Figure 7.8. The rows of M are translates of Y and the columns are translates of X.

assume that X and Y are sorted, for to sort them takes only $O(n \log n)$ time. Then the rows and columns of M are already sorted, and we ask how much more reordering work is necessary until the entire matrix is sorted. The lower bound argument breaks down because the special structure of the matrix permits only rather few permutations.

Claim. The matrices $M = X + Y$ define fewer than n^{8n} permutations.

Proof. For simplicity, consider only matrices with pairwise different entries; that is, $x_i + y_j - x_k - y_\ell \neq 0$ whenever $ij \neq k\ell$. Two pairs of vectors X, Y lead to different permutations if and only if the signs of $x_i + y_j - x_k - y_\ell$ are different for at least one choice of four indices. Think of X, Y as a point in $\mathbb{R}^{2n}$. Then this condition is equivalent to saying that the two pairs correspond to two points on opposite sides of the hyperplane $x_i + y_j - x_k - y_\ell = 0$. There are fewer than n^4 quadruplets of indices to be considered. They correspond to fewer than n^4 hyperplanes that cut up $\mathbb{R}^{2n}$ into fewer than

$$\binom{n^4}{2n} + \binom{n^4}{2n-1} + \cdots + \binom{n^4}{0} < n^{8n}$$

chambers; see, for example, [1]. Each chamber corresponds to a permutation.
□

Michael Fredman [2] showed that there exists a binary decision tree that sorts M in $O(n^2)$ comparisons. However, it is not clear how to construct a path along this tree in $O(n^2)$ time.

Question. Does there exist a comparison-based algorithm that sorts $M = X + Y$ in $O(n^2)$ time?

Steiger and Streinu show that $O(n^2 \log n)$ time suffices to find $O(n^2)$ comparisons for sorting $X + Y$.

[1] H. Edelsbrunner. *Algorithms in Combinatorial Geometry.* Springer-Verlag, Heidelberg, Germany, 1987.

[2] M. L. Fredman. How good is the information theory bound on sorting? *Theoret. Comput. Sci.* **1** (1976), 355–361.

[3] L. H. Harper, T. H. Payne, J. E. Savage, and E. Straus. Sorting $X + Y$. *Comm. ACM* **18** (1975), 347–349.

[4] W. Steiger and I. Streinu. A pseudo-algorithmic separation of lines from pseudo-lines. *Inform. Process. Lett.* **53** (1995), 295–299.

P.8 Union of disks

Let A and B be two sets of n unit disks in the plane. Let $a_1, a_2, \ldots, a_n$ be the centers of the disks in A and $b_1, b_2, \ldots, b_n$ the centers of the disks in B. We call B a *contraction* of A if every pair of disks in B is at least as close as the corresponding pair in A,

$$\|b_i - b_j\| \leq \|a_i - a_j\|,$$

for all i and j. More than 40 years ago Thue Poulsen [6] and independently Kneser [5] asked whether the area shrinks if the disks move closer together.

Question. Is it true that $\text{area} \bigcup B \leq \text{area} \bigcup A$?

It is tempting to conjecture that $\bigcup B$ is indeed smaller or at least not larger than $\bigcup A$, but no proof is currently available. Bollobás [2] proves the conjecture in the special case where there are continuous maps $f_i : [0, 1] \to \mathbb{R}^2$ with

(i) $f_i(0) = a_i$ and $f_i(1) = b_i$, and
(ii) $\|f_i(u) - f_j(u)\| \leq \|f_i(t) - f_j(t)\|$,

for all i and j and all $0 \leq t \leq u \leq 1$. In this case we say there is a *deformation contraction* from A to B. The trouble is that there are contractions B for which there is no deformation contraction, and Figure 7.9 shows a minimal example to that effect. The proof proceeds in two steps. The first step establishes the relation for the perimeter of the union, $\text{per} \bigcup B \leq \text{per} \bigcup A$. The second step observes that the area is the integral of the perimeter over all radii,

$$\text{area} \bigcup A = \int_{r=0}^{1} \text{per} \bigcup A(r) \, dr$$

$$\geq \int_{r=0}^{1} \text{per} \bigcup B(r) \, dr$$

$$= \text{area} \bigcup B,$$

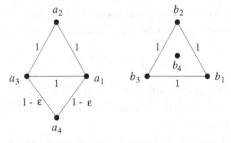

Figure 7.9. Center a_4 moves from outside to inside the equilateral triangle of the other three centers. Its distance to at least one other center increases during a portion of the motion.

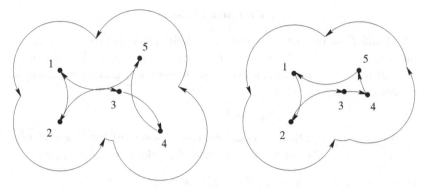

Figure 7.10. From left to right the centers move closer together, which implies that the arcs of the dual perimeter get shorter.

where $A(r)$ is the set of disks with centers a_i and radius r, and similar for $B(r)$. We present a new argument for the first step.

Claim. If there is a deformation contraction from A to B then $\mathrm{per}\bigcup B \leq \mathrm{per}\bigcup A$.

Proof. Imagine we trace the perimeter of $\bigcup A$ with a compass that draws at both tips. We alternate between drawing a circular arc of the boundary and a circular arc connecting two disk centers. If the boundary is connected, as in Figure 7.10, we can draw both curves from beginning to end without lifting either tip. Call the second set of arcs the *dual boundary* and denote its length by $\mathrm{dper}\bigcup A$. The compass turns in a anticlockwise and a clockwise order depending on whether it draws the boundary or its dual. In total it turns 360° relative to its original position, which implies

$$\mathrm{per}\bigcup A - \mathrm{dper}\bigcup A = 2\pi.$$

During the deformation contraction, the centers move closer together and the arcs of the dual boundary can only get shorter. The length difference is constant, which implies that both perimeters can only decrease, and in particular

$$\mathrm{per}\bigcup B \leq \mathrm{per}\bigcup A.$$

The above argument applies only in the case where the boundary is a single curve and the sequence of disks contributing arcs remains the same during the entire deformation contraction.

In general, the boundary consists of one or more curves, namely one per component of the union and one per hole. For each component, the length of the curve minus the length of its dual is 2π, and for each hole it is -2π. The total length difference is $2\pi(\beta_0 - \beta_1)$, where β_0 is the number of components and β_1 is the number of holes. Again the difference is constant so the earlier argument applies. Finally, we remove the restriction to deformation contractions that keep the sequences of disks contributing arcs invariant. We do this by cutting the time interval into discrete segments. Within each segment the sequences are unchanged and the argument applies. Points in time that separate segments correspond to degenerate sets of disks, where either two circles touch or three or more circles pass through a common point. The claim follows because the transition from one segment to the next takes zero time and does not allow for any change in perimeter. □

We note that the proved relation for the perimeter fails for general contractions. An example by Habicht with per $\bigcup B >$ per $\bigcup A$ is described in the open problem book by Klee and Wagon [4] and is illustrated in Figure 7.11. The centers of the disks in A lie on three cocentric circles with radii $R - 1$, R, $R + 1$. The disk centers are fairly dense on the outer circle. For each center on the outer circle there is a corresponding center on the same radiating half-line on the inner circle. The disk centers on the middle circle are fairly sparse, just dense enough to cover the circle. The contraction moves every disk centered on the outer circle straight to its corresponding disk on the inner circle. The perimeter of A consists of two components. The inner component approximates the circle with radius $R - 2$, and the outer component approximates the circle with radius $R + 2$. The perimeter of B consists of the same inner component, but the outer component is now a bumpy approximation of the circle with radius

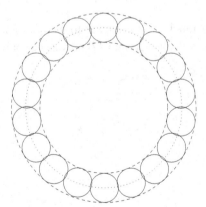

Figure 7.11. The centers of the disks lie on the two dashed and the one dotted circle. Only the disks with centers on the dotted circle are shown.

$R + 1$. We have per $\bigcup B$ > per $\bigcup A$ because the bumpiness adds more to the perimeter than we lose by decreasing the radius of the approximated circle.

Csikós [3] and independently Bern and Sahai [1] prove that under assumption of a deformation contraction, the area relation holds even for unions of disks that are not all the same size. Their arguments are not based on the perimeter, which indeed no longer changes in a monotonous fashion.

[1] M. Bern and A. Sahai. Pushing disks together – the continuous-motion case. *Discrete Comput. Geom.* **20** (1998), 499–514.
[2] B. Bollobás. Area of the union of disks. *Elem. Math.* **23** (1968), 60–61.
[3] B. Csikós. On the Poulsen-Kneser-Hadwiger conjecture. In *Intuitive Geometry*, Bolyai Mathematical Studies **6**, Bolyai Society, Budapest, 1997, 291–300.
[4] V. Klee and S. Wagon. *Old and New Unsolved Problems in Plane Geometry and Number Theory.* Math. Assoc. Amer., Dolcini Math. Exp. **11**, 1991.
[5] M. Kneser. Einige Bemerkungen über das Minkowskische Flächenmass. *Arch. Math.* **6** (1955), 382–390.
[6] E. Thue Poulsen. Problem 10. *Math. Scand.* **2** (1954), 346.

P.9 Intersection of disks

Let A and B be two sets of n unit disks in the plane. As before, we call B a contraction of A if every pair of disks in B is at least as close as the corresponding pair in A, $\|b_i - b_j\| \leq \|a_i - a_j\|$, for all i and j. We are interested in the case where the disks have a nonempty common intersection, as in Figure 7.12. The area of the intersection of two disks increases as the disks move closer together. Can we make a similar statement for any number of disks?

Question. Is it true that area $\bigcap B \geq$ area $\bigcap A$?

It is generally conjectured that the answer to the question is affirmative, but currently no proof is available. Gromov [2] proves the conjecture for three unit

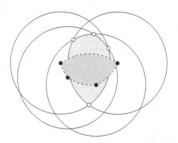

Figure 7.12. Set of four disks, the intersection of the four disks, and the intersection of the dual disks drawn on top of everything else.

disks and its generalization for $d + 1$ unit d-balls in $\mathbb{R}^d$. Capoyleas [1] proves the conjecture in the special case where A and B are connected by a deformation contraction. As for the union, it suffices to prove the inequality for the perimeter. The inequality for the area follows by integrating the perimeter over all radii from 0 to 1.

Claim. If there is a deformation contraction from A to B then $\text{per} \bigcap B \geq \text{per} \bigcap A$.

Proof. Let Z be the set of unit disks with centers in $\bigcap A$. Each disk $z \in Z$ contains all centers of disks in A. As illustrated in Figure 7.12, the boundaries of $\bigcap A$ and of $\bigcap Z$ consist of circular arcs. After central reflection of $\bigcap Z$, we can merge the two sets of arcs to form a unit circle. The length of the circle is

$$\text{per} \bigcap A + \text{per} \bigcap Z = 2\pi.$$

Consider a period of time during which the deformation contraction keeps the sequence of disks contributing arcs invariant. The arcs of $\bigcap Z$ can only get shorter. Since the sum of the two perimeters is constant, this implies that the perimeter of $\bigcap A$ can only increase. □

[1] V. Capoyleas. On the area of the intersection of disks in the plane. *Comput. Geom. Theory Appl.* **6** (1996), 393–396.

[2] M. Gromov. Monotonicity of the volume of intersection of balls. In *Geometrical Aspects of Functional Analysis*, Lecture Notes in Mathematics **1267**, Springer-Verlag, Berlin, 1987, 1–4.

P.10 Space-filling tetrahedra

Given any arbitrary one triangle, we can cover the plane with congruent and nonoverlapping copies of that triangle. We can even lay out the copies such that any two are either disjoint or meet along a common edge or vertex. Such a layout is called a *tiling*, and the triangle is said to *tile* the plane. The situation in $\mathbb{R}^3$ is different; namely, we cannot do the same even with the regular tetrahedron.

We need a few definitions to continue. Two tetrahedra are *congruent* if one can be obtained from the other by an orthogonal transformation, or equivalently by a sequence of translations, rotations, and reflections. A tetrahedron τ *tiles* $\mathbb{R}^3$ if we can cover $\mathbb{R}^3$ with copies of τ such that the intersection of any two copies is either empty or a common triangle, edge, or vertex. Although the regular tetrahedron does not tile $\mathbb{R}^3$, there are other tetrahedra that do. A *space-filling* tetrahedron is one that tiles $\mathbb{R}^3$.

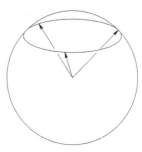

Figure 7.13. Three equally long shape vectors whose endpoints are equally spaced along their circle.

To construct a space-filling tetrahedron τ, we recall the definition of the edgewise subdivision discussed in Section 3.2; see also [1]. Let $\mathbf{v}_1, \mathbf{v}_2, \mathbf{v}_3$ be the shape vectors of τ. The tetrahedra in the subdivision of τ all have the same three shape vectors, but they may come in different orders. These tetrahedra are all congruent if for each permutation (i, j, k) of $(1, 2, 3)$ there is an orthogonal transformation that maps $\mathbf{v}_1$ to $\mathbf{v}_i$, $\mathbf{v}_2$ to $\mathbf{v}_j$, $\mathbf{v}_3$ to $\mathbf{v}_k$. These orthogonal transformations exist if and only if $\|\mathbf{v}_1\| = \|\mathbf{v}_2\| = \|\mathbf{v}_3\|$ and the angles between pairs of vectors are the same. A configuration that satisfies this condition is shown in Figure 7.13. There is a one-parameter family of pairwise noncongruent such configurations, and they are parametrized by the angle between the vectors.

A particularly symmetric tetrahedron in this one-parametric family is defined by $\|\mathbf{v}_1 + \mathbf{v}_2 + \mathbf{v}_3\| = \|\mathbf{v}_1\|$. It is the tetrahedron that arises in the Delaunay triangulation of the body-centered cube (BCC) lattice. As illustrated in Figure 7.14, this lattice is $2\mathbb{Z}^3 \cup [2\mathbb{Z}^3 + (1, 1, 1)]$, and each Delaunay tetrahedron has two vertex disjoint edges of length 2 and the four remaining edges of length $\sqrt{3}$.

Constructions of one-parameter families of space-filling tetrahedra that are different from the one above can be found in Sommerville [3]. However, there is still the open question of characterizing all space-filling tetrahedra [2].

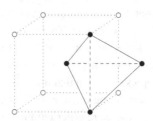

Figure 7.14. Small portion of BCC lattice and its Delaunay triangulation.

Question. Is there a finite collection of rules that characterizes all space-filling tetrahedra?

[1] H. Edelsbrunner and D. R. Grayson. Edgewise subdivision of a simplex. *Discrete Comput. Gemo.* **24** (2000), 707–719.
[2] M. Senechal. Which tetrahedra fill space? *Math. Mag.* **54** (1981), 227–243.
[3] D. M. Y. Sommerville. Space-filling tetrahedra in Euclidean space. *Proc. Edinburgh Math. Soc.* **41** (1923), 49–57.

P.11 Connecting contours

Surfaces are often reconstructed from contour data, which consist of polygons in a sequence of parallel planes in $\mathbb{R}^3$. If we are able to connect polygons in contiguous planes, we can glue the pieces together to form a larger surface representing the data. Meyers, Skinner and Sloan [2] survey algorithms that reconstruct surfaces this way. Let us take a closer look at the problem of connecting two contours.

Let P and Q be polygons in parallel planes in $\mathbb{R}^3$. Connecting P with Q means constructing a triangulated cylinder glued to P on one side and to Q on the other, as illustrated in Figure 7.15. In topological language, the cylinder is a homotopy between P and Q. Each triangle connects an edge of P with a vertex of Q, or vice versa. In either case, it has two edges that lie in neither plane. Along these two edges, the triangle is connected to the predecessor and to the successor around the cylinder.

Let the vertices of P be labelled from 0 to $m-1$ and those of Q from 0 to $n-1$. We use a directed graph laid out on a torus to understand the structure of all possible triangulated cylinders connecting P with Q. The nodes are the pairs $(i, j) \in \text{Vert } P \times \text{Vert } Q$. From each (i, j) there are directed arcs to $(i+1, j)$ and $(i, j+1)$, where $i+1$ is the successor of i in a ccw order around P, and $j+1$ is the successor of j in the same order around Q. The graph is

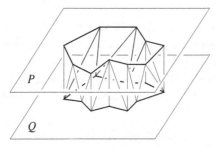

Figure 7.15. Cylindrical connection between two contours.

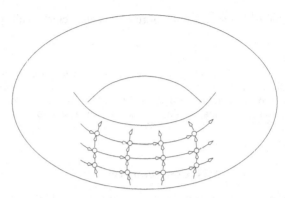

Figure 7.16. Portion of graph representing cylindrical connections between two contours.

illustrated in Figure 7.16. Each edge is a node, each triangle is an arc, and a triangulated cylinder is a directed cycle that winds around once in each of the two torus directions. Following Fuchs, Kedem, and Uselton [1], we search for area minimizing cylinders. The minimum area cylinder among all cylinders that contain a fixed edge can be computed in time O(mn) by using dynamic programming. We let $(0, 0) = (m, n)$ be the fixed edge. For each node (i, j), the algorithm finds the minimum total area of a path/partial cylinder from $(0, 0)$ to (i, j). That area is denoted as $A_{i,j}$. Assume $A_{i,j} = 0.0$ and area $(i, j, k) = 0.0$ whenever one of the indices is negative.

```
for i = 0 to m do
  for j = 0 to n do
    A_{i,j} = min{A_{i-1,j} + area (i − 1, i, j),
                 A_{i,j-1} + area (i, j − 1, j)}
  endfor
endfor.
```

The minimum area of a cylinder containing $(0, 0)$ is $A_{m,n}$. To compute the minimum area without any restricting edge/node, we construct m cycles, one each for $(i, 0)$ for i from 0 to $m − 1$. The resulting running time is O(m^2n), which can be improved to O($mn \log m$) by using the fact that the area minimum cylinder/cycle for $(i, 0)$ lies between those for $(i − 1, 0)$ and $(i + 1, 0)$.

Question. Can the area minimal cylinder connecting P with Q be constructed in time O(mn)?

[1] H. Fuchs, Z. M. Kedem, and S. P. Uselton. Optimal surface reconstruction from planar contours. *Commun. ACM* **20** (1977), 693–702.
[2] D. Meyers, S. Skinner, and K. Sloan. Surfaces from contours. *ACM Trans. Graphics* **11** (1992), 228–258.

P.12 Shellability of 3-balls

Let K be a triangulation of $\mathbb{B}^3$. A *shelling* is an ordering of the d-simplices such that every prefix defines a d-ball. K is *shellable* if it has a shelling. We proved in Section 3.4 that every triangulation of $\mathbb{B}^2$ is shellable. Danaraj and Klee [2] show that such a shelling can be found in time proportional to the number of triangles. The algorithm starts with an arbitrary triangle and adds other triangles greedily. This works because every partial shelling of $\mathbb{B}^2$ can be extended to a complete shelling.

Question. Is there a polynomial time algorithm that decides whether or not a given triangulation of $\mathbb{B}^3$ has a shelling?

For this question to be meaningful, it must be the case that not all triangulations of $\mathbb{B}^3$ are shellable. We describe a nonshellable example shortly. If we use the greedy algorithm, we either succeed in constructing a shelling or we get stuck because none of the remaining tetrahedra can be added to our current ordering. Ziegler [3] shows that this is not an indication for the nonshellability of the 3-complex. Indeed, even three-dimensional Delaunay complexes, which are known to be shellable, can have partial shellings that are not extendable.

The house with two rooms is a nonshellable triangulation of the 3-ball. It is described in the survey paper by Bing [1] and is sketched in Figure 7.17. There are two rooms, one above the other. The lower room is accessible through a

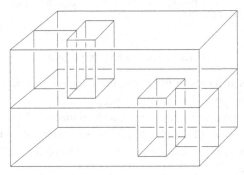

Figure 7.17. House with two rooms. We can construct it from a solid block of clay without tearing or breaking.

chimney passing through the upper room. To avoid a noncontractible cycle in the upper room, we connect the chimney with a screen to the wall. The upper room is accessible through a chimney passing through the lower room. Again we use a screen to avoid a noncontractible cycle. Each wall, floor, ceiling, chimney, and screen is thickened to one layer of cubic bricks. All vertices of a cube belong to the boundary of the house, but edges and squares may belong to the boundary or the interior. We refer to the connected components of faces that belong to the boundary as *exposures* of the cube. For example, a cube in the middle of a wall has two exposures, each consisting of a square and its four edges and four vertices. By construction, every cube in the complex has at least two exposures. In other words, no cubic brick can be last in a shelling of the house. The complex of cubes is not shellable.

To extend the construction from cubic to tetrahedral bricks, we decompose each cube into six tetrahedra. The decomposition has to be consistent at shared squares. This is achieved by using a global ordering of the vertices. Each square has four vertices, and we connect the first vertex in the ordering to the opposite two edges. Each cube has eight vertices, and we connect the first vertex in the ordering to the opposite six triangles. The ordering is constructed by distinguishing three types of vertices. Each vertex belongs to an exposure for each of its cubes, and its *type* is the minimum dimension of any of these exposures. In the ordering, vertices of type 0 precede vertices of type 1, and vertices of type 1 precede vertices of type 2. Because of this rule, every tetrahedron has at least two exposures. Hence the complex of tetrahedra is not shellable either.

[1] R. H. Bing. Some aspects of the topology of 3-manifolds related to the Poincaré conjecture. In *Lectures on Modern Mathematics II*, T. L. Saaty (ed.), Wiley, New York, 1964, 93–128.
[2] G. Danaraj and V. Klee. Which spheres are shellable? In *Algorithmic Aspects of Combinatorics*, B. Alspach et al. (eds.), *Ann. Discrete Math.* **2** (1978), 33–52.
[3] G. M. Ziegler. Shelling polyhedral 3-balls and 4-polytopes. *Discrete Comput. Geom.* **19** (1998), 159–174.

P.13 Counting halving edges

Let S be a set of n points in the plane. For simplicity assume n is even and no three points are collinear. A *halving edge* is an edge $uv \in \binom{S}{2}$ such that the line passing through u and v partitions the remaining $n - 2$ points into equally large sets on both sides of the line. Figure 7.18 illustrates the idea by showing all halving edges of a set of eight points. Let $h(S)$ be the number of halving edges, and define

$$h(n) = \max\{h(S) \mid \text{card } S = n\}$$

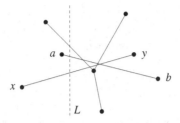

Figure 7.18. Eight points with five halving edges. The dashed line intersects three of them.

for even numbers n. It is clear that $h(n) \leq \binom{n}{2}$, but to improve this trivial bound is not entirely straightforward. The first nontrivial upper bound of $c \cdot n^{3/2}$ was proved 1971 by Laszlo Lovász [3]. In the early 1990s the bound was improved every so slightly by Pach, Steiger, and Szemerédi [4]. The currently best upper bound of $c \cdot n^{4/3}$ is from Tamal Dey [1]. We reconstruct Lovász' proof of the $c \cdot n\sqrt{n}$ bound, which can also be found in [2]. It is based on the following fundamental lemma, which is also used in the proofs of the improved bounds.

Lemma. A line crosses at most $(n+2)/2$ halving edges.

Proof. Let ab, xy be halving edges crossing a vertical line L, as in Figure 7.18. Assume the slope of xy exceeds that of ab. Then we claim that to the left of L there are fewer points above the line ab than above the line xy. If the intersection point of the two lines is to the right of L, as in Figure 7.18, then this is obvious. Otherwise, the reverse is obvious for the points to the right of L, and by the property of halving edges our claim is true to the left of L. Hence, each halving edge crossing L is associated with a different number of points to the left of L and above the line of the edge. We may assume that at most half of the points lie to the left of L, which implies the claimed bound. $\qquad \square$

Draw $n-1$ vertical lines decomposing $\mathbb{R}^2$ into n strips, each containing one of the points. The number of edges that cross $\sqrt{n}$ or fewer of the lines is at most $n\sqrt{n}$. The total number of intersections between halving edges and lines is less than n^2. It follows that the number of halving edges that cross $\sqrt{n}$ or more of the lines is at most $n\sqrt{n}$. This implies $h(n) \leq 2n\sqrt{n}$.

Even though a lot of time and effort was invested in proving upper bounds, it is generally believed that even $n^{4/3}$ is far beyond $h(n)$. The lower bound of $h(n) \geq c \cdot n \log_2 n$ proved in 1973 by Erdős et al. [2] has been improved to $h(n) \geq n \cdot 2^{\sqrt{c \log_2 n}}$ by Géza Tóth [5]. This bound is asymptotically less than $n^{1+\varepsilon} = n \cdot 2^{\varepsilon \log_2 n}$ and asymptotically more than $n \log_2 n = n \cdot 2^{\log_2 \log_2 n}$. Because of the apparent difficulty of the problem, we replace the quest for the asymptotic order of $h(n)$ by a more modest goal.

Question. Is it true that for every $\varepsilon > 0$ there is a constant $c = c(\varepsilon)$ such that $h(n) \le c \cdot n^{1+\varepsilon}$ for every even n?

[1] T. K. Dey. Improved bounds on planar k-sets and related problems. *Discrete Comput. Geom.* **19** (1998), 373–383.
[2] P. Erdős, L. Lovász, A. Simmons, and E. G. Straus. Dissection graphs of planar point sets. In *A Survey of Combinatorial Theory*, J. N. Srivastava et al. (eds.), North-Holland, Amsterdam (1973), 139–149.
[3] L. Lovász. On the number of halving lines. *Ann. Univ. Sci. Budapest Eötvös Sect. Math.* **14** (1971), 107–108.
[4] J. Pach, W. Steiger, and E. Szemerédi. An upper bound on the number of k-sets. *Discrete Comput. Geom.* **7** (1992), 109–123.
[5] G. Tóth. Point sets with many k-sets. *Discrete Comput. Geom.* **26** (2004), 187–194.

P.14 Counting crossing triangles

Suppose S is a set of n points in $\mathbb{R}^2$, and assume for convenience that no three lie on a common line. Two edges of the complete graph defined by S *cross* if they share a point that is interior to both. Ajtai et al. [1] show that if we pick t of the $\binom{n}{2}$ edges, then the number of crossing pairs is at least some constant times t^3/n^2, provided $t \ge 4n$. The lower bound is asymptotically tight. Indeed, we can pick the vertices of a regular n-gon and the t shortest edges connecting the points, as indicated in Figure 7.19. The number of crossing pairs is roughly $n \sum i^2$ with i from 1 to $\lfloor \frac{t}{n} \rfloor$, which is roughly t^3/n^2, as in the lower bound. We

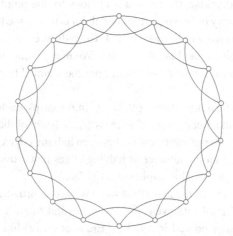

Figure 7.19. Straight edges are drawn as circular arcs to improve the display of crossings.

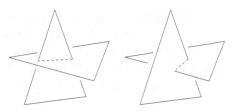

Figure 7.20. Two triangles cross if either one pierces the other or if their boundaries are linked.

extend the counting problem to $\mathbb{R}^3$, where the known lower and upper bounds no longer match asymptotically.

Let now S be a set of n points in general position in $\mathbb{R}^3$, and assume for convenience that no four lie on a common plane. A subset $U \subseteq S$ of three points defines the triangle $\sigma_U = \text{conv } U$. The collection of $\binom{n}{3}$ triangles is denoted as $\binom{S}{3}$. Two triangles *cross* if they intersect without sharing any vertices, $\sigma_U \cap \sigma_V \neq \emptyset$ and $U \cap V = \emptyset$, as shown in Figure 7.20. For a subset of triangles $T \subseteq \binom{S}{3}$, let $x(S, T)$ denote the number of crossing pairs in T and consider the minimum over all sets of n points and t triangles,

$$x(n, t) = \min\{x(S, T) \mid \text{card } S = n, \text{card } T = t\}.$$

With the use of the Euler characteristic for triangulations of the 3-sphere, it is not difficult to prove that the number of crossing-free triangles defined by n points in $\mathbb{R}^3$ cannot exceed $n(n-3)$. There is an example with exactly that many pairwise noncrossing triangles, hence $x(n, t) > 0$ if and only if $t > n(n-3)$. Using ideas from [1], Dey and Edelsbrunner prove there are positive constants c_1, c_2 such that

$$c_1 \cdot t^4/n^6 \le x(n, t) \le c_2 \cdot t^3/n^3$$

whenever $t \geq 2n^2$.

Question. What is the asymptotic order of $x(n, t)$?

The motivation for counting crossing triangle pairs is the problem of counting halving triangles, which are defined by the property that their planes partition S into three points on the plane and the rest in equal halves on each side. It can be shown that a single line cannot cross more than $n^2/8$ halving triangles. Whenever there is a crossing triangle pair, there is an edge of one triangle that crosses the other triangle. Therefore the number of halving triangles cannot exceed the largest t for which

$$c_1 \cdot t^4/n^6 \cdot 3/t \le n^2/8.$$

This value of t is some constant times $n^{8/3}$. It would be nice to increase the lower bound on the number of crossing triangle pairs to a constant times t^3/n^3. This would improve the upper bound on the number of halving triangles to a constant times $n^{5/2}$. That bound was recently established by Sharir, Smorodinsky, and Tardos [3]; however, it did not further our knowledge about $x(n, t)$.

[1] M. Ajtai, V. Chvátal, M. M. Newborn, and E. Szemerédi. Crossing-free subgraphs. *Ann. Discrete Math.* **12** (1982), 9–12.
[2] T. K. Dey and H. Edelsbrunner. Counting triangle crossings and halving planes. *Discrete Comput. Geom.* **12** (1994), 281–289.
[3] M. Sharir, S. Smorodinsky, and G. Tardos. An improved bound for k-sets in three dimensions. In "16th Ann. Sympos. Comput. Geom.," 2000, 43–49.

P.15 Collinear points

Let S be a set of n points in the plane. For each $j \geq 2$, let $C_j(S)$ be the number of collinear j-tuplets. Let $C_j(n)$ be the maximum $C_j(S)$ over all sets of n points without collinear $(j + 1)$-tuplet,

$$C_j(n) = \max\{C_j(S) \mid \text{card } S = n, C_{j+1}(S) = 0\}.$$

For $j = 2$, we can take any set of n points in general position and count $C_2(n) = \binom{n}{2}$ pairs. Determining $C_3(n)$ is known as the orchard problem. Burr, Grünbaum, and Sloane [2] show that

$$C_3(n) \geq 1 + \left\lfloor \frac{n(n-3)}{6} \right\rfloor,$$

which they prove with a fairly involved construction. Füredi and Palásti [3] give a simple example for the same lower bound. The bound almost matches the straightforward upper bound of

$$C_j(n) \leq \binom{n}{2} \Big/ \binom{j}{2} = \frac{n(n-1)}{j(j-1)},$$

which we get by observing that j collinear points account for $\binom{j}{2}$ of the $\binom{n}{2}$ pairs formed by the n points. For $j \geq 4$, the best lower bounds have for a long time been from Branko Grünbaum [4], who shows that

$$C_j(n) \geq c \cdot n^{1 + \frac{1}{j-2}},$$

for some positive constant c that depends on j. It is convenient to describe the lower bound example in the dual plane where each point in S becomes a line. Three points are collinear if and only if the three dual lines meet in a common point. The goal is to construct n lines so that the number of points that belong to j lines is a maximum while at the same time no point belongs to $j + 1$ lines.

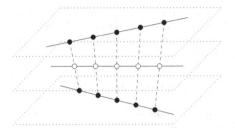

Figure 7.21. Corresponding points of two skew lines are connected by lines, which intersect a plane between the two skew lines in collinear points.

The construction proceeds by induction over j. The solution to the orchard problem is our starting point. For the general step, let A be a set of k lines with

$$p = C_j(k) \geq c \cdot k^{\frac{j-1}{j-2}}$$

points on j lines each. Put A into the plane $x_3 = 1$ in $\mathbb{R}^3$. Let B be a copy of A in the plane $x_3 = \ell$, but rotated by an angle α. We connect corresponding points in A and B, and intersect the p lines with planes $x_3 = i$ for $1 \leq i \leq \ell$, as illustrated in Figure 7.21. Collinear points in $x_3 = 1$ correspond to the same number of collinear points in any plane $x_3 = i$. Why? In each plane $x_3 = i$ we draw k lines such that each intersection point lies in j of these lines. We have a total of $n = p + \ell \cdot k$ lines and $\ell \cdot p$ points on $j + 1$ lines each. Project the lines into a plane and choose α such that no $j + 2$ lines pass through a common point. Finally, choose $\ell = k^{\frac{1}{j-2}}$ and get $n = c' \cdot k^{1 + \frac{1}{j-2}}$ lines and $c \cdot k^{1 + \frac{2}{j-2}}$ points on $j + 1$ lines each. The number of such points is some constant times $n^{1 + \frac{1}{j-1}}$.

A slight improvement of the lower bound for $C_j(n)$ to $c \cdot n^{\log_2(j + 4/\log_2 j)}$, for $4 \leq j \leq 18$, and to $c \cdot n^{1 + 1/(j - 3.59)}$, for $18 \leq j$, is described in [5], but see also [1]. Both in Grünbaum's and in Ismailescu's construction the exponent goes to one as j goes to infinity. Still, there is no upper bound known that for constant j is asymptotically less than n^2.

Question. For constant $j \geq 4$, are there positive constants ε and c such that $C_j(n) \leq c \cdot n^{2 - \varepsilon}$?

[1] P. Braß. On point sets with no k collinear points. In *Discrete Geometry: in Honor of W. Kuperberg's 60th Birthday*, A. Bezdek (ed.), Marcel Dekker, New York, 2003, 185–192.

[2] S. A. Burr, B. Grünbaum, and N. J. A. Sloane. The orchard problem. *Geom. Dedicata* **2** (1974), 397–424.

[3] Z. Füredi and I. Palásti. Arrangements of lines with a large number of triangles. *Proc. Amer. Math. Soc.* **92** (1984), 561–566.

[4] B. Grünbaum. New views of some old questions of combinatorial geometry. *Atti Accad. Naz. Lincei* **17**: *Theorie Combinatorie* (1976), 451–468.
[5] D. Ismailescu. Restricted point configurations with many collinear *k*-tuples. *Discrete Comput. Geom.*, to appear.

P.16 Developing polytopes

A 3-*polytope* is the convex hull of a finite set of points in $\mathbb{R}^3$ that do not all lie in a common plane. It is a convex polytope whose boundary complex consists of facets, edges, and vertices connected like a 2-sphere. Each facet is a convex polygon. After cutting along a spanning tree of the 1-skeleton, the boundary is still connected and can be laid out flat. We call the result a *net* of the 3-polytope. Figure 7.22 illustrates that even rather simple convex polytopes have more than one net. The concept of a net was described hundreds of years ago by the German artist Albrecht Dürer [4]. Of course, when we develop a boundary complex into a net, it might happen that some of the faces overlap. If the net is nonoverlapping, we can construct the polytope from paper by essentially following the inverse procedure: cutting the net out of paper, folding at the edges, and gluing along matching edges of the boundary. The question whether or not every 3-polytope can be made of paper this way is mentioned in the open problem collection by Croft, Falconer, and Guy [3, B21].

Question. Does every 3-polytope have a nonoverlapping net?

Nets of more complicated polytopes than the cube can be found in the design book by Critchlow [2]. Aronov and O'Rourke consider a similar but different question. They study the star unfolding of a convex polytope, which is obtained by cutting the boundary along edges of the shortest path tree that connects an arbitrary point on the boundary with all polytope vertices. As proved in [1], the star unfolding does not overlap. The problem is different because shortest paths

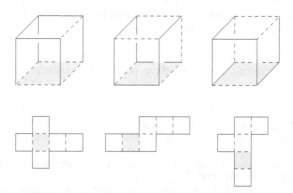

Figure 7.22. We cut along the solid cube edges, which become boundary edges of the net.

generally pass through facets and are therefore not necessarily contained in the 1-skeleton.

Various heuristics for finding nonoverlapping nets have been studied, including minimum spanning trees and shortest path trees restricted to the 1-skeleton. However, for both heuristics there are convex polytopes that lead to overlapping nets.

[1] B. Aronov and J. O'Rourke. Nonoverlap of the star unfolding. *Discrete Comput. Geom.* **3** (1992), 219–250.
[2] K. Critchlow. *Order in Space. A Design Source Book.* Thames and Hudson, New York, 1987.
[3] H. T. Croft, L. J. Falconer, and R. K. Guy. *Unsolved Problems in Geometry.* Springer-Verlag, New York, 1991.
[4] A. Dürer. *Unterweysung der Messung mit Zyrkel und Rychtscheydt.* 1525.

P.17 Inverting unfoldings

An *unfolding* of (the boundary of) a three-dimensional convex polytope is similar to a net, except cuts can also go through facets and the unfolding is assumed to be nonoverlapping. Examples are the star unfoldings mentioned in the preceding open problem on nets, which exist for all 3-polytopes. We consider the problem of inverting the process by forming creases along interior edges of the polygon and gluing matching boundary edges. We assume we have a complete description of the gluing pattern but no information on where the crease edges ought to be. As an example, consider the two rectangular polygons in Figure 7.23, which both can be glued to form a tetrahedron. Because of the different aspect ratios, the layout of the crease edges is different in the two cases. The basic result on folding polygons into 3-polytopes is the possibly surprising theorem by A. D. Alexandrov [1].

Theorem. Let P be a polygon with boundary gluing such that

(1) the angle at every point is at most 2π,
(2) after gluing, P is homeomorphic to $\mathbb{S}^2$.

Then P is the unfolding of the boundary of a unique 3-polytope.

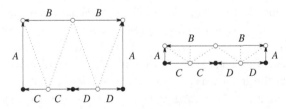

Figure 7.23. Only the hollow vertices have angles less than 2π. The dotted lines are not given but rather implied by the gluing pattern along the boundary.

Using this theorem, we can easily check whether or not a polygon with gluing pattern is the unfolding of a 3-polytope. Assuming it is, Nikolai Dolbilin asked in 1995 how the unique 3-polytope can be reconstructed.

Question. Is there an algorithm that constructs a 3-polytope from its unfolding?

To be specific, there are numerical approximation algorithms [3], but it is not known whether or not the reconstruction can be done exactly. The difficulty is not the lack of knowledge where the crease edges are. Indeed, they are a subset of the graph of shortest paths between vertex pairs [4]. This graph is finite and all possible choices of crease edges can be enumerated in finite time. The difficulty in finding an algorithm is making Cauchy's Rigidity Theorem constructive. It states that up to congruence a 3-polytope is determined by its facets and the adjacencies between them [2]. A deformation of a 3-polytope thus necessarily deforms at least one of the facets.

[1] A. D. Alexandrov. *Konvexe Polyeder.* Akademie Verlag, 1958.
[2] P. R. Cromwell. *Polyhedra.* Cambridge Univ. Press, England, 1997.
[3] J. O'Rourke. Folding and unfolding in computational geometry. In "Proc. Japan Conf. Discrete Comput. Geom.," Lecture Notes in Comput. Sci., Springer-Verlag, 1998.
[4] E. Demaine, M. Demaine, A. Lubiw, J. O'Rourke, and I. Pashchenko. Metamorphosis of the cube. In "Proc. 15th Ann. Sympos. Comput. Geom.," 1999, 409–410.

P.18 Flip graph connectivity

In the plane, every triangulation of a finite point set can be transformed into the Delaunay triangulation by a sequence of edges flips. There is an algorithm that finds a sequence of edge flips whose length is at most quadratic in the number of points. It follows that every triangulation μ of the point set can be transformed into every other triangulation ν by a sequence of quadratically many edge flips: move from μ to the Delaunay triangulation and then to ν.

The situation is more complicated in $\mathbb{R}^3$. We introduce definitions needed to formalize the problem. Let S be a finite set of points in $\mathbb{R}^3$. We let N be the collection of simplicial complexes K with vertex set Vert $K = S$ and underlying space $|K| = \text{conv } S$. The complexes in N are the nodes of the *flip graph* of S and there is an arc between nodes μ and ν if there is a two-to-three flip or a three-to-two flip that takes μ to ν. Such a flip is illustrated in Figure 7.24. In the plane, the arcs correspond to two-to-two flips. What we said above can now be rephrased. The flip graph of a point set in the plane is connected and its diameter is at most quadratic in the number of points. Barry Joe [3] and independently Edelsbrunner, Preparata, and West [2] ask whether a similar claim can be made in three dimensions.

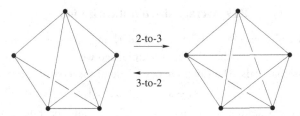

Figure 7.24. A two-to-three flip and its inverse correspond to the two directions we can traverse an arc in the flip graph.

Question. Is the flip graph of a finite set $S \subseteq \mathbb{R}^3$ connected if no four points in S are coplanar?

The restriction to generic point sets is necessary. Take, for example, the set of six vertices of the regular octahedron. As illustrated in Figure 7.25, each tetrahedrization consists of four tetrahedra surrounding one of the three space diagonals. None of the three tetrahedrizations permits the application of a two-to-three or a three-to-two flip. The flip graph thus consists of three isolated nodes.

The current knowledge for generic point sets is rather modest. The answer to the above question is known to be affirmative if the number of points is at most seven. For sets larger than that, it is not even known whether or not the flip graph can have isolated nodes. Such nodes would correspond to tetrahedrizations that do not permit any flip. If we permit tetrahedrizations that use only subsets of the points as vertices, and among these we restrict ourselves to weighted Delaunay tetrahedrizations, then the answer to the question is affirmative. The flip graph they define is the 1-skeleton of a high-dimensional convex polytope known as the fiber and also the secondary polytope [1].

[1] L. J. Billera and B. Sturmfels. Fiber polytopes. *Ann. of Math.* **135** (1992), 527–549.
[2] H. Edelsbrunner, F. P. Preparata, and D. B. West. Tetrahedrizing point sets in three dimensions. *J. Symbolic Comput.* **10** (1990), 335–347.
[3] B. Joe. Three-dimensional triangulations from local transformations. *SIAM J. Sci. Statist. Comput.* **10** (1989), 718–741.

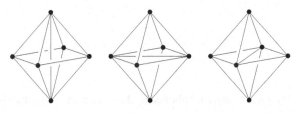

Figure 7.25. The three isolated tetrahedrizations of the six vertices of the regular octahedron.

P.19 Average size tetrahedrization

Consider a cell complex in $\mathbb{R}^3$ whose underlying space is contractible. We naturally assume that each cell is contractible, and we let s_k denote the number of k-dimensional cells. The Euler characteristic of this complex is

$$\chi = s_0 - s_1 + s_2 - s_3 = 1.$$

If the complex is simplicial, we have $4s_3 \leq 2s_2$ because each tetrahedron has four triangles and each triangle belongs to at most two tetrahedra. Write $s_0 = n$ and substitute the inequality into the equation for χ to get $2s_1 - s_2 \geq 2n - 2$. There are $s_1 \leq \binom{n}{2}$ edges, because there are only that many point pairs. This implies

$$s_1 \leq n(n-1)/2,$$

$$s_2 \leq (n-2)(n-1),$$

$$s_3 \leq (n-2)(n-1)/2.$$

In summary, a simplicial complex with n vertices in $\mathbb{R}^3$ has size at most quadratic in n. This bound applies also to Delaunay tetrahedrizations. In many cases, the number of Delaunay simplices is much less than quadratic. For example, Rex Dwyer [2] proves that if the n points are chosen randomly from the uniform distribution in the unit cube then the expected number of edges, triangles, tetrahedra is in $O(n)$. Less is known about other distributions. A rather special open question considers points on a saddle surface as illustrated in Figure 7.26.

Question. What is the expected number of edges in the Delaunay tetrahedrization of n points randomly chosen from the uniform distribution on the hyperbolic paraboloid $z = x^2 - y^2$ inside $[-1, +1]^3$?

Attali, Boissonnat and Lieutier prove that under some reasonable assumptions, the Delaunay tetrahedrization of a set of n points on a smooth surface has at

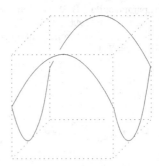

Figure 7.26. The saddle surface is indicated by the four parabolas in which it meets the boundary of the cube.

most some constant times $n \log_2 n$ simplices [1]. The proof requires the surface be generic and the points be more or less evenly sampled. Specifically, there is a constant $\varepsilon > 0$ such that for every ball of radius ε around a point of the surface the number of sampled points inside the ball is at least one and at most some constant. This bound is perhaps the yet strongest piece of evidence that constructing Delaunay tetrahedrizations for the purpose of reconstructing a surface from a finite point sample is not as impractical as the quadratic bound may suggest.

Instead of asking the size question for randomly chosen points, we can impose restrictions on the distribution. For a finite set $S \subseteq \mathbb{R}^3$, let d and D be the minimum distance and the maximum distance between any two points in S. We call $\Delta = D/d$ the *spread* of S. For n points in $\mathbb{R}^3$, the smallest possible spread is a constant times $\sqrt[3]{n}$, and if the lower bound is obtained then the Delaunay tetrahedrization has at most $O(n)$ simplices. Furthermore, all known examples of quadratic size Delaunay tetrahedrizations have points lined up along curves, and such sets have spread at least some constant times n. Two-dimensional distributions, such as the one on the saddle surface, have spread at least some constant times $\sqrt{n}$. Jeff Erickson proves that for all n there is a set of n points with spread $\Delta \leq n$ whose Delaunay tetrahedrization has more than some constant times $n \Delta$ simplices [3]. He also proves that the number of simplices is less than some constant times Δ^3, provided $\Delta \leq \sqrt{n}$ [4]. It would be interesting to close the remaining gap.

[1] D. Attali, J.-D. Boissonnat and A. Lieutier. Complexity of the Delaunay triangulation of points on surfaces: the smooth case. In "Proc. 19th Ann. Sympos. Comput. Geom.," 2003, 201–210.
[2] R. A. Dwyer. Average-case analysis of algorithms for convex hulls and Voronoi diagrams. Ph.D. thesis, Report CMU-CS-88-132, Carnegie-Mellon Univ., Pittsburgh, Pennsylvania, 1988.
[3] J. Erickson. Nice point sets can have nasty Delaunay triangulations. In "Proc. 17th Ann. Sympos. Comput. Geom.," 2001, 96–105.
[4] J. Erickson. Dense point sets have sparse Delaunay triangulations. In "Proc. 13th Ann. ACM-SIAM Sympos. Discrete Alg.," 2002, 125–134.

P.20 Equipartition in four dimensions

Let S be a set of n points in the plane. There exist two lines h_1 and h_2 so each quadrant defined by the lines contains $n/4$ or fewer points in its interior. To illustrate such a 4-partition, Figure 7.27 draws the finite set as a region in the plane. Here is the sketch of a constructive proof that such two lines indeed exist. First, construct h_1 so it cuts S in half. To be precise, let S_-, S_0, and S_+ be the subsets of points on the negative side, on, and on the positive side of h_1. We require that S_- and S_+ each contain at most half the points in S. Points on h_1 are not counted. Note that we can prescribe the direction of h_1. Second, construct a line h_2 that

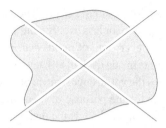

Figure 7.27. Every set in the plane can be cut by two straight lines into four pieces, each at most a quarter of the original size.

cuts S_- in half. If we are lucky then h_2 also cuts S_+ in half and we are done. Otherwise, we rotate h_2 while making sure it always cuts S_- in half. The rotation pivots about points on the side of S_-, so the line sweeps over S_+ without backtracking. This implies that there is a moment in time where h_2 cuts both sets in half.

We generalize the partition problem to d dimensions. Let S be a set of n points in $\mathbb{R}^d$, and let $h_1, h_2, \ldots, h_d$ be d hyperplanes. In the nondegenerate case, the hyperplanes meet at a unique point and decompose $\mathbb{R}^d$ into 2^d orthants. The hyperplanes form an *equipartition* if each orthant contains $n/2^d$ or fewer points in its interior. As argued above, every set in $\mathbb{R}^2$ has an equipartition. In fact, there is one degree of freedom left in the construction, which we can use to prescribe the direction of the first line, or to enforce that the two lines are perpendicular. In 1966, Hugo Hadwiger extended this result to three dimensions by proving that every set in $\mathbb{R}^3$ has an equipartition [2]. There are two leftover degrees of freedom, which he uses to prescribe the normal direction of the first plane.

The situation is different in five and higher dimensions. Avis shows that there are sets in $\mathbb{R}^5$ that do not have equipartitions [1]. We count degrees of freedom to see that this negative result is plausible. A hyperplane in $\mathbb{R}^d$ has d degrees of freedom, and since we have d hyperplanes, we have a total of d^2 degrees at our disposal. If we specify the hyperplanes in sequence, we use a degree of freedom for each set we cut in half. The total number of consumed degrees is $1 + 2 + \cdots + 2^{d-1} = 2^d - 1$. The smallest number of dimensions where we consume more degrees of freedom than we have is $d = 5$. For $d = 4$ we have 16 degrees and need 15. This suggests that an equipartition exists, but no proof is known.

Question. Does every finite set in $\mathbb{R}^4$ have an equipartition?

A host of related results can be found in a paper by Edgar Ramos [3]. He generalizes the Borsuk–Ulam Theorem from topology and proves among other things that every finite set $S \subseteq \mathbb{R}^4$ has an equipartition by four 3-spheres.

[1] D. Avis. Non-partitionable point sets. *Inform. Process. Lett.* **19** (1984), 125–129.

[2] H. Hadwiger. Simultane Vierteilung zweier Körper. *Arch. Math. (Basel)* **17** (1966), 274–278.
[3] E. A. Ramos. Equipartition of mass distributions by hyperplanes. *Discrete Comput. Geom.* **15** (1996), 147–167.

P.21 Embedding in space

Planar graphs are one-dimensional complexes that can be drawn in the plane without crossing. Kuratowski [3] proves that a graph is planar if and only if it contains no subgraph homeomorphic to K_5 or to $K_{3,3}$. Fáry [1] shows that every planar graph has a straight-line embedding in $\mathbb{R}^2$. Hopcroft and Tarjan [2] demonstrate that time proportional to the number of vertices is sufficient to decide planarity on a conventional random access machine. None of these nice results seems to generalize even to three dimensions. We begin with some definitions. An *embedding* of a topological space $\mathbb{X}$ in another such space $\mathbb{Y}$ is an injection $j : \mathbb{X} \to \mathbb{Y}$ whose restriction to the image $j(\mathbb{X})$ is a homeomorphism. A *planar graph* is a 1-complex K with an embedding $j : |K| \to \mathbb{R}^2$.

Every 1-complex has an embedding in $\mathbb{R}^3$, but not every 2-complex does. Even simple 2-complexes such as the Klein bottle cannot be embedded in $\mathbb{R}^3$, but if we cut out one disk it could. Similarly, cutting out a disk from any nonorientable 2-manifold changes the status from nonembeddable to embeddable. There is an infinite sequence of different such 2-manifolds, which implies that there is no finite collection of obstructions that could play the role of K_5 and $K_{2,3}$ for 2-complexes. Fáry's theorem also does not generalize. The construction of a 2-complex that can be embedded but not geometrically realized in $\mathbb{R}^3$ uses a trefoil knot, like the one illustrated in Figure 7.28. Any view of the knot has at least three cross-over points. A polygonal cycle forming a trefoil knot in $\mathbb{R}^3$ certainly consists of more than three edges. To get the 2-complex, we tetrahedrize a box and remove from this tetrahedrization a tunnel in the shape of a trefoil knot. Then we insert a cycle of three (curved) edges and repair the tetrahedrization by connecting the cycle to the tunnel boundary. The 2-skeleton of

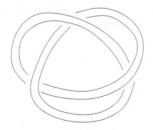

Figure 7.28. A tube in the shape of a trefoil knot in space.

this tetrahedrization has a combinatorially unique embedding but no geometric realization in $\mathbb{R}^3$.

The question of recognizing 2-complexes K with embeddings in $\mathbb{R}^3$ seems difficult. We can construct a system of polynomial inequalities such that K has a geometric realization in $\mathbb{R}^3$ if and only if the system is satisfiable by reals. Tarski's quantifier elimination method effectively determines the satisfiability of such systems. It follows that the recognition of 2-complexes with geometric realizations in $\mathbb{R}^3$ is decidable. Can this result be generalized to embeddings?

Question. Is the recognition of 2-complexes that have embeddings in $\mathbb{R}^3$ decidable?

[1] I. Fáry. On straight line representation of planar graphs. *Acta Sci. Math. (Szeged)* **11** (1948), 229–233.
[2] J. E. Hopcroft and R. E. Tarjan. Efficient planarity testing. *J. ACM* **21** (1974), 549–568.
[3] K. Kuratowski. Sur le problème des courbes en topologie. *Fund. Math.* **15** (1930), 271–283.
[4] A. Tarski. *A Decision Method for Elementary Algebra and Geometry.* Second edition, Univ. California Press, 1951.

P.22 Conforming tetrahedrization

Let K be a 1-complex in $\mathbb{R}^2$. A *conforming Delaunay triangulation* is a Delaunay triangulation D that contains a subdivision of K as a subcomplex. In other words, every vertex of K is a vertex of D and every edge of K is either an edge in D or cut into two or more edges in D. Examples similar to the one in Figure 7.29 can be used to show that the smallest conforming Delaunay triangulation sometimes has at least some constant times n^2 vertices, where n is the number of edges and vertices in K. Edelsbrunner and Tan [1] prove that some constant times n^3 vertices are always sufficient. It is not known whether any of the two bounds is tight or the answer lies somewhere in between.

We generalize the problem to three dimensions and ask two questions, one for 1- and the other for 2-complexes. Let K be a simplicial complex in $\mathbb{R}^3$.

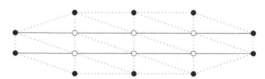

Figure 7.29. A subdivision of the solid 1-complex is subcomplex of the partially dotted Delaunay triangulation.

A *conforming Delaunay tetrahedrization* is a Delaunay tetrahedrization that contains a subdivision of K as a subcomplex.

Question. Does there exist a polynomial function $f(n)$ such that every 1-complex K with n vertices and edges in $\mathbb{R}^3$ has a conforming Delaunay tetrahedrization with $f(n)$ or fewer vertices?

The author remembers proving an exponential upper bound for $f(n)$ years ago. The proof of the existence of $f(n)$ does not extend to 2-complexes. It is conceivable that there are 2-complexes in $\mathbb{R}^3$ for which the number of vertices necessary for a conforming Delaunay tetrahedrization is a function not only of size but also of relative distance between the simplices.

Question. Does there exist a function $g(n)$ such that every 2-complex K with n vertices, edges, and triangles in $\mathbb{R}^3$ has a conforming Delaunay triangulation with $g(n)$ or fewer vertices?

Call an edge *strongly Delaunay* if there is a sphere that passes through the endpoints and all other vertices lie strictly outside that sphere. Shewchuk [2] proves that if every edge of a 2-complex is strongly Delaunay, then there exists a constrained Delaunay tetrahedrization for that 2-complex. The definition of that tetrahedrization is a generalization of the two-dimensional notion of a constrained Delaunay triangulation discussed in Section 2.1. This result suggests that an affirmative answer to the first question could be a useful tool in mesh generation, even if the answer to the second question turns out to be negative.

[1] H. Edelsbrunner and T. S. Tan. An upper bound for conforming Delaunay triangulations. *Discrete Comput. Geom.* **10** (1993), 197–213.
[2] J. R. Shewchuk. A condition guaranteeing the existence of higher-dimensional constrained Delaunay triangulations. In " Proc. 14th Ann. Sympos. Comput. Geom.," 1998, 76–85.

P.23 Hexahedral mesh size

The number of tetrahedra in a tetrahedral mesh can be as big as quadratic in the number of vertices. For a simple example of this kind, choose n points each on two skew lines and construct $(n - 1)^2$ tetrahedra by connecting every contiguous pair on one with every contiguous pair on the other line, as illustrated in Figure 5.15. Other types of meshes do not permit such disproportionally large numbers of three-dimensional cells. Possibly the most popular because most regular type is the *structured mesh*, which is isomorphic to a subcomplex of the regular cube tiling of $\mathbb{R}^3$. Every three-dimensional cells is a *hexahedron* with face structure isomorphic to that of a three-dimensional cube. Every hexahedron

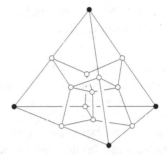

Figure 7.30. Subdivision of a tetrahedron into four hexahedra.

has eight vertices, and every vertex belongs to eight or fewer hexahedra. It follows that the number of hexahedra is at most the number of vertices.

Marshall Bern and David Eppstein ask whether a similarly small upper bound on the number of hexahedra holds independent of the regularity of their connections. A *hexahedral mesh* is a complex in $\mathbb{R}^3$ whose three-dimensional cells are hexahedra. It is no longer required that the hexahedra meet in four around interior edges and in eight around interior vertices. Structured meshes are rather rigid, but hexahedral meshes are just as flexible as tetrahedral meshes. In particular, every tetrahedron can be cut into four hexahedra, as shown in Figure 7.30. Similarly, every tetrahedral mesh can be subdivided into a hexahedral mesh of four times as many hexahedra as there are tetrahedra. At the same time, the number of vertices increases beyond the original number of tetrahedra. In the resulting hexahedral mesh, the number of hexahedra is at most some constant times the number of vertices.

Question. What is the asymptotic order of the maximum number of hexahedra in a hexahedral mesh with n vertices?

The question makes sense in a geometric setting, where each hexahedron is required to be a convex polyhedron, as well as in a topological setting, where each hexahedron is a three-dimensional cell with specific pattern of neighboring cells. The answer to the question might be different in the two settings. Joswig and Ziegler [1] establish a lower bound of a constant times $n \log_2 n$ that applies in both settings. They prove that for large enough d, there are convex 4-polytopes whose 1-skeletons are isomorphic to the 1-skeleton of the d-dimensional cube. The 1-skeleton has $2^d = n$ vertices and $d \cdot 2^{d-1} = n/2$ $\log_2 n$ edges. The Schlegel diagram of such a 4-polytope is a hexahedral mesh with at least as many hexahedra as there are edges.

[1] M. Joswig and G. M. Ziegler. Neighborly cubical polytopes. *Discrete Comput. Geom.* **24** (2000), 325–344.

Subject Index

Author Index